BY HAKEEM OLUSEYI

A Quantum Life
A Quantum Life (Adapted for Young Adults)

WHY DO WE EXIST?

WHY DO WE EXIST?

THE NINE REALMS OF THE UNIVERSE
THAT MAKE YOU POSSIBLE

HAKEEM OLUSEYI

WITH NILS JOHNSON-SHELTON

BALLANTINE BOOKS

NEW YORK

Image credits are listed on page 231.

Hardcover ISBN 978-1-984-81912-3
Ebook ISBN 978-1-984-81913-0

Printed in the United States of America on acid-free paper

1st Printing

FIRST EDITION

BOOK TEAM: Production editor: Cassie Gitkin • Managing editor:
Pamela Alders • Production manager: Meghan O'Leary •
Proofreaders: Liz Carbonell, Rachael Clements, Julie Ehlers, Jinah Yoon

Book design by Simon M. Sullivan

The authorized representative in the EU for product safety and
compliance is Penguin Random House Ireland, Morrison Chambers,
32 Nassau Street, Dublin D02 YH68, Ireland.
https://eu-contact.penguin.ie

*For every curious mind that has ever looked up
at the night sky and whispered:
Why?*

CONTENTS

WHY DO WE EXIST?

INTRODUCTION

I'll never forget the first time I observed a distant galaxy with only my eyes.

I was in Northern California at a friend's house, and I'd brought a pair of small telescopes with me to the coastal mountains near the town of Elk. My friend and I peered skyward side by side, playing a game of pointing the telescopes at no particular location to see if we could find something interesting. At one point, my friend, who was not a physicist and had never looked through a telescope before that night, said, "Bruh, what's that fuzzy thing? I can see it even without the scope."

I didn't believe him. The only naked-eye fuzzy blob I could think of was the Great Nebula in Orion, twenty-four light-years across and around thirteen hundred light-years away. But the key word here is "Orion." We weren't looking in Orion, which was in the east that time of year and low on the horizon. We were looking north, high in the sky toward Cassiopeia. It didn't add up.

I went to his telescope, peered through the eyepiece, then moved my eye away and looked along the telescope barrel to the sky. And there it was. "What the heck is that?" I blurted.

Then it hit me: We were looking at the Andromeda galaxy, the nearest large galaxy to the Milky Way, a member of our Local Group. Although I had often viewed Andromeda at observatories and on computer screens, I had never seen it like this:

with my own eyes, through Earth's atmosphere, from a dark mountaintop.

As I stared upward, the sky suddenly transformed from a two-dimensional "celestial dome" to a three-dimensional vista spanning millions of light-years. Before me was the Milky Way's plane of gas, dust, and stars, stretching some 87,000 light-years across. And far beyond, at 2.5 million light-years more distant, loomed the Andromeda galaxy, which I knew to be almost twice the size of the Milky Way at 150,000 light-years across. Modern humans did not even exist when the light falling on my retinas began its trip across the expanse of space and time.

Objects look smaller the farther away they are, and yet this thing—this hazy blob—was huge: six times the width of the full Moon in our sky. My mind reeled. I could feel the scale in my bones. For the first time, I truly saw and felt the staggering vastness of the Cosmological Realm, where galaxies and galaxy clusters reign.

I stood next to my friend and explained in my excited, profane way just how mother-effing huge that thing was. (I swear a lot—but not in this book—publisher's orders!)

I imagined myself as a cosmic Marvel Universe superbeing, a giant towering over the galaxies, scooping up Andromeda and the Milky Way like tadpoles from a stream. What would it be like to wade through the universe like a colossus in a galactic river?

Catastrophic, probably! I'd implode into an ultra-supermassive black hole, tearing through space-time, devouring galaxies, and ripping open a wormhole to another universe.

But I digress.

That night sparked a new way of seeing for me. I began to imagine reality as layered—not just in scale, but in rules. The realm of galaxies—what I call the Cosmological Realm—is governed by different laws from the one we live in day to day, which I call the Middle Realm. Still different are the rules of the Quantum Realm, governing particles far too small to see. Each realm operates under distinct principles, but somehow they all coexist, fitting together like interlocking gears. I didn't just realize this; I imagined it, conjured it, birthed it. The Nine Realms are my Scientific Wild-Ass Guess—my SWAG—for organizing reality.

This isn't a textbook. It's not even a standard pop-sci tour of physics. It's a SWAG at the biggest question of all: Why do we exist? This book is my attempt to weave together what we know, what we suspect, and what we can barely imagine, through the lens of the Nine Realms that make your existence possible. Each chapter explores a distinct domain of the universe governed by its own laws, scales, and deceptions. These realms are scientific, but they're also psychological, philosophical, and imaginative. They're grounded in physics, but not constrained by it. Together, they form a new map of reality that puts you, the reader, at its center.

I'll give you the facts, yes, but also the doubts, the weirdness, the poetry, and hopefully show what it's like to ask some of these bold questions without flinching.

This book is about what happens when we bring imagination and observation together. That's what physicists do more than anything: We watch. Chemists mix and manipulate. Biologists dissect and culture. Geologists hike and hammer. But physicists? We observe. Then we imagine the rules behind what we see, converting this into the language of math.

Yes, we use machines—giant particle accelerators and telescopes—but those are just extensions of our eyes. We make observations, form SWAGs, and refine them through testing. If the universe says "nope," we move on to the next guess.

One of the first to operate this way was Ibn al-Haytham, an Iraqi polymath working in the eleventh century. He became obsessed with how the eye works, and in 1021 CE he published *The Book of Optics*, a seven-volume scientific treatise grounded in one radical idea: that only observation should determine what we accept to be true.

This sounds obvious now (to scientists, at least). But at the time, it was heretical. It meant truth wasn't absolute or ordained by scripture; it was conditional, revisable. That was dangerous thinking.

Later, others would pay the price for holding similar beliefs. Galileo was punished by the church for saying Earth revolved around the Sun. Giordano Bruno was burned at the stake for proposing stars were distant suns with their own planets. And in my own time, I caught flak for publishing the truth that James Webb—the NASA administrator whose name graces our newest space telescope—was not a bigot, despite popular accusations. Speaking truth, even when backed by evidence, can still get you burned.

And yet, those obsessed with truth keep pushing. Galileo didn't invent the telescope, a Dutchman did, but the famous Italian did make improvements to it. He was one of the first to point it at the stars, not just the horizon. My friend and I had used a slightly more advanced version of the same contraption to fall into Andromeda on that fateful California night.

Those first telescopes weren't much more than brass tubes with a couple of glass lenses at either end. Over time, our instruments have gotten better. We eventually realized we could analyze light and its spectrum to discern what a thing was made of, and that the universe made noises, and that everything that was something radiated energy—so we built things that could detect spectra, and radio waves, and infrared radiation. We built mountaintop observatories and vast arrays that span the globe to listen to our cosmos. Now we have imagined and engineered space telescopes like the Hubble and the James Webb, parked one million miles from Earth at a point called Lagrange 2, where the gravitational effects of Earth and the Sun cancel out. These devices have helped us discover and refine what is true about the universe.

The results? Over the last 125 years, we've learned more

about the universe than in all of human history before it. What will we know in 2100? In 3000? In 10000? What will everyday citizens of those eras take for granted that we don't yet comprehend?

We are on the cusp of a new frontier in science, one where we can now tell a compelling story about our universe and its meaning. We have learned so much and have so many data that we can begin to answer questions like "How much intelligent life is in the universe?" "Why does time exist?" "When did time start?" "How is energy created?" "What is energy?" "What is the purpose of life?" "Why does the universe exist?"

"Why do we exist?"

Scientists are usually squeamish about these types of questions. But I say science is ready to start answering some of them. We'll never reach the final answers, but just as the arc of the moral universe bends toward justice, the arc of inquiry bends toward greater knowledge. And whether you're a physicist, a grocer, a parent, or a child, this is a voyage we take together.

At times, our journey may seem bleak. By volume, nearly the entire universe is hostile to life. It continually evolves to destroy all matter. The distant future holds nothing but oblivion, except for maybe a handful of supermassive black holes. There will be no hope.

Fortunately, this bleak future is really, really, really, really, really, really far away. It means that for us, in our universe's relative infancy, there *is* hope. To me, that hope is represented by energy and imagination. I say that energy is what makes everything happen (sounds logical, but there are other people who think it's information or even disorder). As long as energy is flowing, there's possibility.

Our universe's energetic activities play out on an incomprehensibly massive canvas. The Milky Way alone has hundreds of billions of stars—Andromeda has around a trillion—and the observable universe has hundreds of billions of galaxies. Cosmologically speaking, we live on a grain of sand on an endless beach. Believing that any specific life-form, planet, star, galaxy, or universe is exceptional is as absurd as picking a single grain from our endless beach and claiming it to be the most important one.

But here's the thing about big numbers: They make the possible certain. The rare becomes the inevitable. Yes, humans and our remarkable brains are unlikely. Earth's extremely uncommon life-supporting characteristics have thus far not been found elsewhere. Yet, despite the odds, here we are. The Nine Realms help us see how. Each realm reveals a unique aspect of how reality works and contributes to the conditions that make life—and you—possible.

In this book, we'll travel through these realms: the Middle Realm, the Realm of Life, the Cosmological Realm, the Dark Realm, the Quantum Realm, the Temporal Realm, the Multiverse Realm, the Realms Beyond Horizons, and the Realm of Imagination. The Nine Realms are my SWAGgy attempt to organize reality into something graspable and useful for everyone, not just nerds with astro PhDs.

Before we really get going, I'll make a few promises: no math overload; your mind will be blown; and you will finish this book. Okay, I can't make that last promise with all the things a book has to compete with, from video games to TikTok to making dinner, but I'm pretty sure you will.

If you came here expecting dry science writing, I've got bad news: I have jokes. I have metaphors. I have opinions. I'll use comic book references and quantum superposition in the same breath. Because that's how the universe talks to me. And I think maybe it talks to you that way, too.

Scientists are loath to admit it, but ultimately our SWAGs are products of our imaginations. Someone had to imagine light as a particle and then show how it acted like one. Someone else had to imagine light as a wave and show how it acted like one, too. Each had to define "particle" and "wave" in these instances, since these are made-up ideas about the world, ones we solidly take for granted but first had to have been proven to exist.

The same principle applies to the universe, from the most minute scale to the largest. In this sense, we live in the universe, but we also imagine it into comprehensibility.

Let's get on our California mountaintop, peer through that scope, and dream of how we can splash some stars.

MIDDLE REALM

It ain't what you don't know that gets you into trouble. It's what you know for sure that just ain't so.

—credited to Mark Twain (but we ain't sure if that's so)

Humans used to believe Earth—and humanity—was at the center of everything. Then we learned that Earth orbits the Sun, which orbits the center of the Milky Way, which just kind of floats weightlessly among a swarm of hundreds of billions of other galaxies connected by a colossal invisible superstructure.

The universe definitely does not revolve around us. We were mistaken.

But here's the thing about the universe: Size matters. And while we may not reside at the universe's center, we are in the middle of it—at least in terms of scale.

Our starting realm is the one in which we live—the only realm where life is possible—a place I call the Middle Realm. In terms of scale, the Middle Realm includes everything larger than an atom to the full length and breadth of a galaxy's arm. I call it the Middle Realm not to be cheeky or to suggest that it is "just right" for life (by volume, practically all of the Middle Realm is uninhabitable), but because humans are in the middle of the largest and smallest known physical distances in the entire observable universe.*

* The logarithmic middle!

The greatest perceptible distance is the size of the observable universe itself, which is 10^{26} meters across (a 1 followed by 26 zeros). The least perceptible distance is the size of the smallest known physical entity, a subatomic particle called the neutrino, which measures 10^{-26} meters (a 1 preceded by 25 zeros and a decimal point). By chance, humans are at the scale of 10^0 meters, slapdab in the middle!

The most critical structures of the Middle Realm are stars and their gigantic nurseries. The most essential building block of the Middle Realm is the humble hydrogen molecule, H_2. Neither you nor I would exist without these components.

How are they related, and how do they come together to create things like solar systems and asteroid belts and moons? And how the heck do they produce a planet like Earth, the only planet we know of that is capable of hosting complex, intelligent life? We will answer these questions as we journey from the largest to the most minuscule structures of the Middle Realm, and eventually to us.

■ ■ ■

Before we get started, we have to overcome a problem.

Most of us hold incorrect assumptions about how the universe works that can hinder our understanding of the Nine Realms. So, let's start by deprogramming our minds to overcome what I call normal deceptions. These deceptions aren't lies told with bad intent; rather, they are misunderstandings we pick up simply by being human. They arise from intuition, language, cultural assumptions, our everyday observations, schoolyard myths, even well-meaning teachers. The kicker? They feel true. That's what makes them so effective.

These deceptions influence our understanding of how the universe functions. They are not merely trivial errors; they cause us to overlook complexity, misinterpret cause and effect, and mistakenly believe that things are separate when they are, in fact, deeply interconnected. By actively addressing these deceptions, we can enhance our comprehension of the universe and appreciate the intricate relationships that define it.

Before we map the physics of everyday life, let's pull back the curtain on a few everyday normal deceptions. Not because you need to memorize them, but to show you how deep the rabbit hole goes. Part of this book's mission is to dismantle these illusions, because only by seeing clearly can we start asking the right questions—like why do we exist?

Let's begin by looking at a science story everyone knows: Isaac Newton watching an apple fall from a tree. It is a common enough experience—things fall all the time. But Isaac—one of the greatest scientific minds the world has ever seen—had questions: Why does the apple fall? Why does it fall straight down? Why doesn't it float away? Why doesn't it go sideways?

Isaac eventually concluded that some invisible, all-encompassing force must be acting on the apple, pulling it straight to the ground and, once settled, keeping it there. He called this force gravity. He further concluded that gravity acted on all objects on the planet—including the apple tree, Isaac himself, his chair, and his quill. He also intuited that gravity is the same force that keeps the Moon orbiting Earth and both of these orbiting the Sun.

This tale has stood the test of time because it is relatable. You don't need to be a scientist to understand what Isaac was getting at. Our experience of reality—our normal deception—neatly conforms with the fact that things fall (and often break). Releasing an object from some height and observing it travel toward the ground is as commonplace and reliable an earthly phenomenon as the Sun rising. Falling objects and our rising Sun are not strange at all. They are normal. Yet they are both deceptions.

It would be strange if I released an object and it just hovered above the ground. If that happened, how would you react?

Shock? Wonder? A pinch of dread? Maybe you'd go into full-on freak-out mode? I probably would. It would mean something is not right.

But here's the thing: Falling isn't normal. Not on a cosmic scale. Watching an apple plummet to the ground might feel mundane, but it's actually weird. The real question is this: Why does it fall at all?

Let's reframe Newton's apple moment. Imagine an apple tree floating in deep space. If an apple detaches, it doesn't fall; it just hangs there, motionless. In most places in the universe, there's nothing nearby to pull it in. And without a nearby object, the default state is coasting, drifting, doing nothing. If stillness and constancy are the norm, then accelerated motion—the kind that happens to the apple—should be what surprises us.

This brings us to something that trips up many people: orbiting. A satellite circling Earth isn't flying like a plane; it's falling. Constantly. But it's also moving sideways so fast that, as it falls, the ground curves away beneath it at the same rate. That's what orbiting really is: falling with just enough horizontal motion to keep missing Earth.

Here's the part people may miss: An orbiting satellite is weightless, just like the people inside. So is everything else that's falling. That includes Earth, the Moon, the Sun—heck, the entire Milky Way. If you've ever wondered, "What does Earth weigh?"—well, it's in free fall around the Sun, so the answer is nothing. Weight is what you feel when something resists your fall. But when everything is falling together, there's nothing to push back.

So Chicken Little was right. The sky is falling!

Let's take another look at Isaac's falling apple. Our everyday intuition tells us that once the apple is free from its branch, the distance between it and the ground will decrease at an ever-increasing rate. The apple is decreasing its potential energy (a measure of its energy relative to its position to the ground). At the same time, it increases its kinetic energy (a measure of its energy associated with movement as it hurtles toward the ground). We call this falling rate the gravitational acceleration of Earth. Some of you may remember this constant from high school physics—9.8 meters per second squared. But how do you distinguish between the apple falling to the ground and the ground pushing up toward the apple?

This may sound like a ridiculous question. Clearly, our planet isn't expanding outward to meet all falling objects, from a feather to the Moon. More apparent still, the apple doesn't sit in place while the ground nudges toward it. Yet, our modern understanding of gravity maintains that the ground does, in fact, accelerate up toward the apple.

Here's why. About two hundred years after Newton, another genius came along and refined our knowledge of gravity: Albert Einstein. In Einstein's general theory of relativity—general relativity, or GR for short—gravity is seen not as a force pulling objects down but as the curvature of space-time caused by mass and energy. This leads to a different interpretation of what it means to "fall."

If that sounds perplexing, don't worry, I'll explain space-time in detail in the Cosmological Realm. All you need to know for now is that whenever an object is released anywhere in the universe, it follows a straight path through space-time. We call this path a geodesic. If space-time is curved, as it is near massive objects like Earth, the geodesic appears to us as "falling." The object isn't experiencing a force; it's simply following the most natural, "straight" path through curved space-time.

A somewhat familiar analogue for this geodesic can be found in children's museums and science centers worldwide. (Places that always need funding, by the way!) You've probably seen one of those coin funnels where you drop a quarter in and it rolls around in an ever-tightening spiral before vanishing into a hole at the bottom. The quarter isn't actively "seeking" the hole; it's simply moving along a curved surface shaped by the funnel. Its motion appears circular because of the shape of the surface it's rolling along.

Something similar happens with space-time. Anything in the universe possessing mass and energy curves space-time around it, much like the shape of the funnel. The more massive and energetic the object, the more space-time curves around it. Any other object moving in this curved space-time follows what seems to us like a curved or falling trajectory, even though it's actually moving in a straight line through the distorted geometry.

From the perspective of general relativity, an object in free fall—like an apple falling from a tree or a moon orbiting a planet—is following a natural trajectory determined by space-time. Meanwhile, an object at rest on the ground—like Newton's famous apple or you sitting in your chair reading this book—is *not* following this natural path. Instead, the ground interferes with this natural trajectory, pushing up against, say, you, and preventing you from following your free-fall geodesic around the center of Earth. In this sense, the ground exerts a force upward on you. It only gives the impression that gravity is "pulling" you down; in reality, the ground accelerates upward relative to your natural geodesic in curved space-time. This is how physicists understand gravity to work.

The reason this idea seems counterintuitive is that we often associate acceleration with forward motion. When we hear "the ground accelerates upward," we envision Earth's surface physically expanding outward. However, that's not what's actually occurring. I set you up and exploited that assumption when I guided you in that initial direction. Acceleration isn't solely about increasing speed; a change in direction also constitutes a form of acceleration, which requires a force.

Consider two examples. First, imagine swinging a yo-yo around in a circle at the end of its string. At every instant, the yo-yo accelerates toward your hand—this is centripetal acceleration—because the string exerts a tension force pulling it inward. However, the yo-yo doesn't actually move closer to your hand; instead, it follows a circular path. Without the string, the yo-yo would fly off in a straight line.

A second, more relevant example is the classic carnival ride where riders stand against a spinning wall. As the ride speeds up, the floor drops away, but the riders remain pinned to the wall. Why? Because their bodies want to move in a straight line, but the wall forces them into a circular path. The wall continuously pushes inward on the riders, preventing them from following their natural trajectory.

Now, imagine the riders have no idea they are on a spinning ride. Suppose they were extremely small relative to the structure. All they would know is that they feel "stuck" to the wall; it feels like their "ground." They might conclude that some invisible force is pulling them toward the wall. In reality, the wall accelerating inward keeps them from following their natural geodesic.

This is similar to what happens in curved space-time. Motion through space-time can be thought of as moving in two perpendicular directions: through space and through time. If an object is "at rest" and alone in deep space, far from any massive bodies, it "hovers" and its only space-time motion is its progression through time. However, if we place it above the surface of a massive body, it will start to fall. It cannot hover. Why? Because space-time curvature has altered its natural path, *redirecting part of its motion through time into motion through space*. The object is not pulled down; its direction through space-time is altered—just like the yo-yo on a string or the riders in the carnival ride. The curvature causes the acceleration.

This is what Einstein revealed: Gravity is not a force in the Newtonian sense; it is space-time's geometry and everything's inability to be at rest relative to space-time. The apple doesn't fall down. Earth's surface is curving up in space-time to meet the apple.

How's that for some deprogramming?

Here's the thing: We all carry mental models of how nature is supposed to behave. We inherit them from experience, from school, from culture, and much of the time they're wrong. Not because we're stupid, but because the universe isn't built for intuition. It's built on laws that don't care whether we "get" them.

As we go forward, keep this in mind: Many of the things you think you know about motion, time, space, and even cause and effect are shaped by the biases of the Middle Realm—our narrow little bubble of perception. And this book is a kind of deprogramming.

Across the Nine Realms of reality—Middle, Life, Cosmological, Dark, Quantum, Temporal, Multiverse, Beyond Horizons, and Imagination—we're going to pull apart the assumptions that keep us blind. Then we will build a clear picture of how the universe actually works. Not just for the sake of knowledge, but because understanding what's real might just change what we think is possible.

■ ■ ■

Now that we've cleared out some of the most common misconceptions—our "normal deceptions"—we're ready to see a more accurate picture of how our world came to be. That means starting with the basics: the particles and forces that shape the Middle Realm. We'll begin with the smallest players—electrons—then zoom out to the stars that forged the elements, Earth that collected them, and finally, the emergence of life itself.

Remember, the main currency of the universe is energy. Here in the Middle Realm, the flow of energy is dominated by large masses such as planets and stars, as well as tiny, indivisible particles of matter called electrons.

If I asked a random high school student to define "electron," they might say something like "a negatively charged particle in an atom," and they wouldn't be wrong. But electrons are so much more than that. In many fundamental ways, electrons and their behaviors are the basis for all ordinary matter in the universe. (By ordinary matter, I mean physical matter that we can easily perceive using our senses.) Because of the inherent relationship between mass and energy, electrons are also the basis for all common energy in the universe. And not just that! They are also the basis for light.

You may now be thinking, "Whachoo talking about, Dr. O-lis? Electrons are responsible for matter, energy, *and* light? Why has no one told me that before?" I don't actually know. It's not a conspiracy. Chalk it up to our normal deception—and some difficult-to-update teaching traditions.

Let's first consider electrons and their relationship to matter. As you may already know, atoms consist of a dense nucleus surrounded by a cloud of "orbiting" electrons (to be clear, whatever electrons do around nuclei, they definitely do not constantly fall toward them as satellites do around Earth). The particles that make up an atom's nucleus—the baryons we call protons and neutrons—weigh almost two thousand times more than electrons. If an electron weighed one pound, protons and neutrons would weigh nearly a ton. Put another way, if an electron were a loaf of bread, a proton would be an adult male polar bear or a small car.

This difference in mass has guided our perception of atoms and biased our understanding of reality. In the same way that we view giant planets like Jupiter and Saturn as having captured some of their moons (which is true), we have been taught to think of the much more massive atomic nucleus as "capturing" the lighter electrons. This is our normal deception at work.

But what if the opposite were true? What if electrons were the ones that captured the nuclei? This is like asking, "What if Europa captured Jupiter?" or "What if Newton's apple captured Earth?" Preposterous questions. Isaac Newton certainly never considered them.

Except when it comes to electrons, they aren't preposterous. Consider what our universe would look like if electrons did not exist. Since all protons have the same electric charge, and like charges repel each other, if there were no electrons, every proton in the universe would try to get as far away as possible from every other proton. There would be no clumps of matter considerably larger than a proton. That would be a pretty dull universe!

The mere existence of electrons changes everything.

The fact is that electrons *do* capture protons. And once our tiny electron captures its much larger proton, the repulsive force between protons is shielded and effectively annulled. In this electron-encircled state, our single proton and single electron team up to become an electrically neutral hydrogen atom.

Hydrogen, which accounts for approximately 93 percent of all atoms, is by far the universe's most common element. (No hydrogen would mean no water, and since you and I are roughly 70 percent water, no you or me, either.) With our protons' mutual repulsion eliminated, hydrogen atoms can now gather in unfathomably large numbers.

And they do! The smallest components of the Middle Realm are hydrogen molecules—two bonded hydrogen atoms, or H_2— but the largest individual structures of the Middle Realm are giant clouds of hydrogen molecules, clouds that are ten million times the mass of our Sun and up to six hundred light-years across. We call these giant molecular clouds (GMCs). They serve as the sole source material for new stars. Without electrons, there would be no GMCs, and without GMCs, there would be no stars or planets.

Now let's look at electrons as they relate to energy and light. When protons and electrons are hot and moving very fast in each other's presence (they are virtually always in each other's presence), the electrons can't capture them. This state is called plasma, the universe's most common state of matter. (As you may recall from chemistry class, the other common states are gas, liquid, and solid.)

Electrons can only capture protons once their temperature falls below about 5,000°F (about 3,000° Kelvin)—which to you or me is still crazy hot! At these "lower" temperatures, electrons do everything possible to prevent protons from gaining energy and heating back up. Once an electron manages to capture a proton, it acts like a sort of buffer, absorbing the energy of all the commotion going on around the new hydrogen atom. When the electron absorbs excess energy in one of these interactions, it immediately dumps this extra energy by converting it to light, which streams across the universe in all directions.

But electrons don't just capture protons, cancel their charge, and create light.

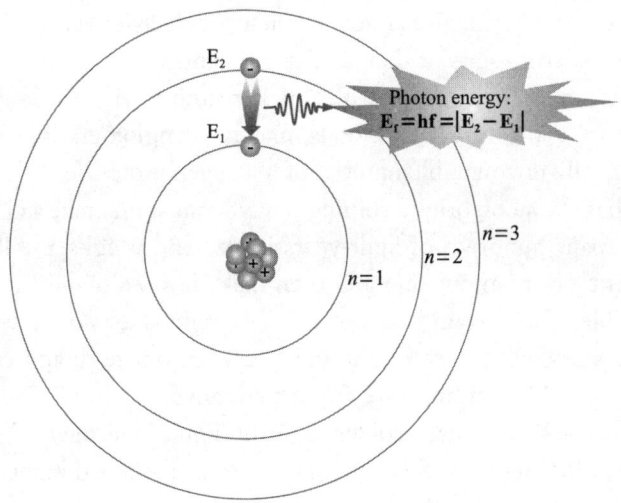

Fig. 1: Atom's Electron Emitting Light
An electron has been excited to energy level 2 (n = 2) by some process like heat absorption. The electron sheds this energy in the form of light that possesses an amount of energy equal to the difference between energy levels 2 and 1.

They also create life.

Our normal deception states that electrons are just tiny negatively charged particles. But in reality, they are the indisputable foundation of the Middle Realm. The universe's electrons are scurrying around, capturing protons, keeping them at lower energy, and gathering them together to form molecules and larger structures—stars, crystals, minerals, planets, and, believe it or not, you.

To our knowledge, complex life requires a relatively small Middle Realm object known as a planet. But planets don't just pop out of nowhere (unless, maybe, wormholes are involved—but that's for much later in this book). To have a planet of any kind, let alone one suitable for life, we first need a star. To create all the elements required for a rocky planet like Earth, we need a whole multigenerational family of deceased stars, for without these we won't have relatively heavy elements like iron, nickel, silicon, or oxygen, and almost all elements heavier than helium originate from stars or their stellar remnants.

Fortunately, the Middle Realm is teeming with those star-birthing giant molecular clouds, massive conglomerations of a practically uncountable number of hydrogen molecules.

To make a hot, bright, shining star, you must first take a GMC that spans hundreds of light-years in size and weighs a million times more than our Sun and then make it as cold and dark as possible. Once again, our normal deception does not serve us well. Stars, which are bright and hot, start out dark and cold. And by cold, I mean more frigid than anything you'll find on Earth: −440°F! That's colder than the liquid hydrogen NASA uses to fuel its giant Space Launch System rockets designed to return astronauts to the Moon.

To this point, we have been discussing electrons and their role in forming hydrogen and GMCs. But to get to stars, we have to transition from electrons back to gravity. Our GMC star nursery has to be cold so that gravity can work its magic, eventually unlocking the GMC's abundant nuclear mass energy, which is locked away inside its protons.

Here's how it plays out. Once our GMC reaches a sufficiently low temperature, density variations occur, forming a weblike network of filaments within the cloud. These filaments are characterized by enhanced density and extend light-years in distance. Eventually, the most densely concentrated filaments become turbulent and unstable, causing them to break apart into a series of clumps. These clumps are typically about a hundred times more massive than the Sun and span a few light-years in diameter.

Over the course of tens of thousands to hundreds of thousands of years, these clumps accumulate more mass by gravitationally attracting additional material or colliding with other clumps. As they grow denser, the clumps break apart and collapse into cores. The cores are approximately ten times the mass of the Sun and several thousand times larger than our solar system. Eventually, some of these cores will become stars.

These cores, which are hundreds of thousands of times larger than the stars they will eventually form, are diffuse and extended. As gravity continues to drive their collapse, the innermost 25–50 percent of the core undergoes intense gravitational compression, forming a roughly spherical protostar. Meanwhile, the surrounding material flattens into a rotating protoplanetary disk. This process lays the groundwork for the formation of a solar system.

In the heart of the swirling core, the protostar teeters on the edge of becoming a true star. Its fate is uncertain because as gravity relentlessly pulls the core inward, it compresses the gas, increasing both its density and its temperature.

It's a common misconception to equate temperature with heat. More accurately, temperature measures the average speed of particles in any given solid, liquid, gas, or plasma. If an object's particles move quickly, it's hot; if they're creeping along, it's cold.

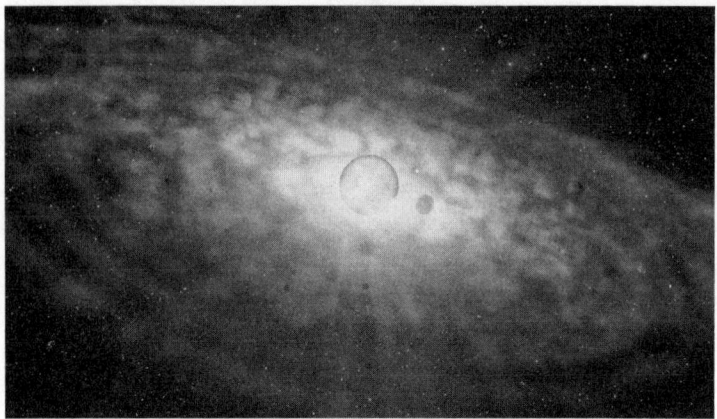

Fɪɢ. 2: Sᴛᴀʀ ᴀɴᴅ Pʟᴀɴᴇᴛᴀʀʏ Sʏsᴛᴇᴍ Fᴏʀᴍᴀᴛɪᴏɴ
Stages of star and planet formation: from giant molecular cloud (GMC) to filamentary collapse, protostar, and protoplanetary disk formation, leading to a mature planetary system.

In this sense, a thermometer functions like a speedometer for particle motions.

The core's rising temperature increases its internal pressure, counteracting gravity's compression. As the particles move faster, they collide more often and with greater force, which ramps up the pressure. The core's pressure is further amplified by its gravitational contraction as the same number of hydrogen atoms are squeezed into an ever-shrinking volume. If the core can't find some way to deal with these pressure increases, it will never compress enough to ignite hydrogen fusion and achieve true stardom. And most don't. Protostars inside a GMC frequently fail to make the leap. Estimates vary, but for every star that makes it, there are between one and one hundred protostars that fail to reach full stellar status. A typical GMC can produce anywhere from tens of thousands to millions of stars, which translates to hundreds of millions of failed stars that are scattered across our galaxy like cosmic false starts.

As if this weren't bad enough, it gets worse for our aspirational protostar.

You see, our core rotates. Cores at this stage always rotate; there will always be turbulence within and around the protostar at the center of the core. Within this turbulence, more particles will be moving in one direction than in any other. This primary direction of motion will eventually win out, causing the rotation of the entire protostar to begin and then accelerate. Just trust me: Nonrotating cores ain't happening.

And because of a universal law of nature known as conservation of angular momentum, the smaller the protostar becomes, the faster it rotates. (This is precisely the phenomenon figure skaters take advantage of when they spin on the ice. As they pull their arms in to their centerlines, they spin faster and faster.

Fig. 3: Binary Star System
Two stars gravitationally bound in mutual orbit—an elegant solution to the angular momentum barrier that impedes star formation.

When they extend their arms out again, the rotation slows.) The faster our protostar rotates, the greater the centrifugal effect counteracting gravity's compression.

If our protostar is ever going to succeed at compressing enough to ignite nuclear fusion in its core and become a star, then this won't do. Fortunately, the laws of physics are not totally aligned against our protostar.

Our protostar deals with the rotation problem by splitting into two or more smaller protostars, almost always of differing mass. When it does so, the rotational energy gets redirected from the rotations of a single protostar into the orbital motion of two protostars dancing around each other. Stable configurations of two stars are called binary stars or, more colloquially, double stars.

The night sky bears witness to the importance of this cloud-splitting process. Half the stars visible to the naked eye from Earth are not single but double stars—two stars locked in orbit around each other.

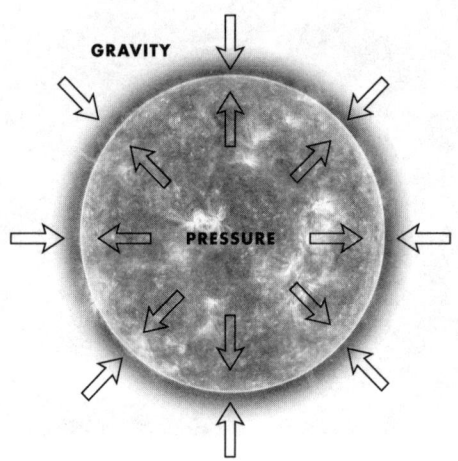

FIG. 4: SUN'S HYDROSTATIC BALANCE
How the Sun maintains balance over time between gravitational contraction and outward pressure from nuclear fusion in its core.

Single stars like our Sun still form, though. While binaries solve the spin problem by giving each other something to sling angular momentum into, single stars outsource the job to their surroundings: magnetic fields, jets, winds, and the protoplanetary disk itself.

With our rotation problem taken care of, let's go back to pressure. We can't do anything about our protostar getting smaller—it will have to get a lot smaller to become a star—but we can do something about the heat that is building up inside.

Exactly how does our shrinking, heating, spinning protostar lose enough of its heat energy to keep going?

The same way you do, Superstar! It uses its electrons.

Right now, and every moment you're alive, you are emitting infrared light. All opaque matter (matter that light cannot freely pass through) emits light. The wavelength of the light it emits depends on its temperature. The protostar's surface now has a temperature of around 1,000° Kelvin, while the temperature at its center is more than a thousand times hotter (although it is still not hot enough to trigger nuclear fusion). One thousand degrees Kelvin may sound hot, but as far as the universe is concerned, this is still quite cool, meaning our protostar emits relatively low-energy infrared light. By contrast, the surface of the Sun, which emits visible light (plus a lot of other higher-energy radiation), is closer to 6,000° Kelvin.

In virtually all instances, electrons are responsible for converting heat energy into light, including infrared light. This is precisely what happens in our protostar. It "shines." It's a dark, black object to the naked eye but a bright, shining object to our infrared telescopes.

Here, we see a concrete example of energy being the universe's currency. If our protostar couldn't use its electrons to transfer heat energy away from it, then no star would ever be able to form. Luckily, though, this energy transfer does occur, and as more and more infrared light is shed, the protostar's atoms slow down just enough to regulate its pressure. Ultimately, this cooling allows the protostar's atoms to succumb, at long last, to gravity.

Once our protostar shrinks past a threshold, its fate is sealed. It begins to spin faster and heats more, yet it continues to shrink. At this stage, the protostar's collapse into a star cannot be stopped. Nuclear fusion ignites once the central temperature reaches about 10 million degrees Kelvin. This is when protons and their kin have no choice but to come together to form the nuclei of elements like helium, carbon, and oxygen (and later, through other stellar processes, almost every other naturally occurring element in existence).

It's helpful to think of our new star as possessing two main parts: the core, which lies at its center, and the surrounding envelope of matter. Take our Sun: Its core occupies the innermost 25 percent of the Sun's diameter but contains 50 percent of its mass; the other half of its mass is in the envelope. The core generates energy, and the envelope responds. The energy created by nuclear fusion pours outward from the core. This energy continually "blows up" the star like a balloon, providing an internal pressure that prevents the star from collapsing further.

The greater the star's mass, the greater the gravitational force

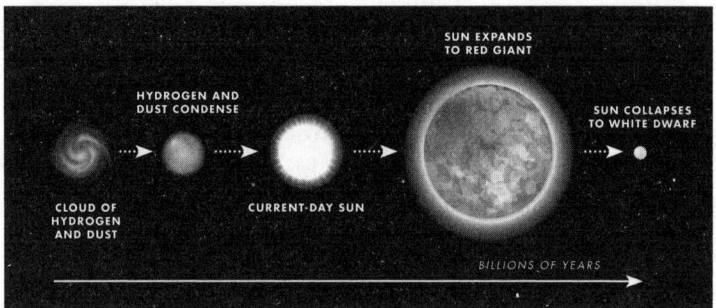

Fig. 5: Sun Size Evolution
The Sun's size expands dramatically across its lifetime—from main sequence to red giant—driven by changes in core fusion rates and energy output.

pushing inward toward the core. This requires increasing rates of core nuclear fusion to produce enough outward radiation pressure to support the envelope's weight. Consider the differences between the cores of the most massive hydrogen-burning stars, called blue giants, and the least massive ones, called red dwarfs. Blue giants are up to a hundred times the Sun's mass, have core temperatures reaching 100 million degrees Kelvin, and are about ten thousand times brighter than the Sun. Red dwarfs are usually less than half the Sun's mass, possess core temperatures ranging only up to 10 million degrees Kelvin, and output only about 1 percent of the Sun's brightness.

A blue giant's core fuses hydrogen at rates millions to billions of times faster than a red dwarf's core. A red dwarf, despite its slow fusion rate, is also vastly more fuel efficient, leading to lifetimes that exceed the current age of the universe, while blue giants burn bright and fast, lasting only a few million years.

Over the star's life, from the moment of birth until its death, the core temperature only increases, continually boosting the star's energy output. This relationship is exponential. If a star doubles its core temperature, its energy output goes up one hundred thousand times. The star's envelope responds by expanding, providing the star with a larger surface area to radiate away this extra energy.

When the Sun initiated core fusion 4.6 billion years ago, the core temperature was about 10 million degrees Kelvin. Today, the Sun's core is 15 million degrees Kelvin. Not incidentally, this means that the Sun was 30 percent dimmer when it formed than it is today.

Stars create virtually all elements larger than helium, from the carbon in our bodies to the gold in our jewelry. Physicists call this process stellar nucleosynthesis. It's a process that starts from the moment a star begins core fusion and continues throughout its life—and beyond.

For most of its life, a star is busy fusing hydrogen into helium; this is nuclear fusion. For stars like our Sun, this happens through

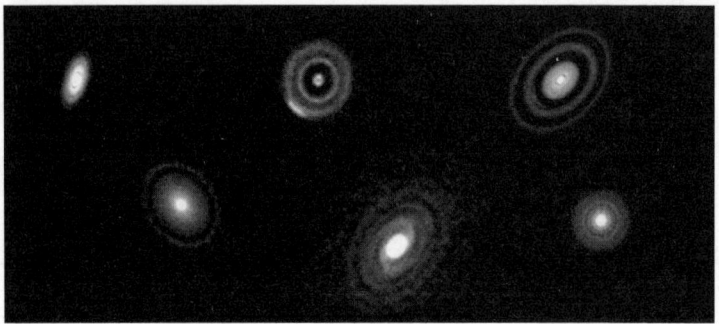

FIG. 6: PROTOPLANETARY DISKS
Flattened disks of gas and dust surrounding young stars, the birthplaces of planets, moons, and other solar system bodies.

something called the proton-proton chain, where four hydrogen nuclei (protons) are mashed together to create a helium nucleus. This releases a lot of energy, which powers the star and results in the light emitted from a star's surface—thank you, electrons!

In larger, hotter stars, there's a more complex process for doing the same thing called the CNO cycle, where carbon, nitrogen, and oxygen act like helpers, speeding up the hydrogen fusion process.

Eventually, the star exhausts the hydrogen in its core, and things become interesting. The star then shifts to helium fusion, which lasts only a tenth as long as hydrogen fusion. This reaction is known as the triple-alpha process, in which three helium nuclei combine to form a carbon nucleus. If the star is massive enough, it can go further, fusing carbon into oxygen, neon, magnesium, and even heavier elements. The catch is that every time the star moves to fuse a heavier element, it requires much higher temperatures and pressures in its core. Additionally, the durations of these heavier element fusion cycles become shorter and shorter, meaning the window to make these elements gets smaller and smaller.

For example, a ten solar mass star will fuse hydrogen for about ten million years, helium for about one million years, carbon for about ten thousand years, neon for about one year, and silicon for just a single day. The result of silicon fusion is iron. For stars, iron fusion is a cosmic brick wall. Fusing iron doesn't release energy; it consumes it. No star that consumes its own energy is going to last very long.

Not all stars are massive enough to keep fusing elements up to iron. Smaller stars, such as our Sun, undergo a quieter phase known as the asymptotic giant branch (AGB), during which they burn helium and hydrogen in shells surrounding an inert carbon-oxygen core. In this phase, the star is able to fuse elements heavier than carbon and oxygen, only not in the core, but in its envelope. This reaction is known as the *s*-process, or slow neutron capture.

Nuclei in the star's envelope capture neutrons one by one, and those nuclei subsequently become unstable. Every element has a sweet spot for the ideal number of neutrons. If the number of neutrons in a nucleus is greater than or less than the ideal, neutrons or protons will transmute into each other by processes known as beta decay—where neutrons transmute and become protons—or reverse beta decay, where protons transmute and become neutrons. In the *s*-process, beta decay dominates, and the star's envelope slowly builds up heavier elements like strontium, barium, and lead. These elements eventually get mixed up to the star's surface and blown out into space by the stellar wind. In the end, AGB stars—which our Sun will be in five to seven billion years—gently shed their outer layers, enriching the galaxy with these heavier elements.

But the fate of the most massive stars is far more dramatic. Once they hit that iron wall, the core becomes unstable and collapses, triggering a supernova explosion. During this explosion, things happen so fast that the star doesn't just capture neutrons slowly as in the *s*-process. Instead, it undergoes the *r*-process, or rapid neutron capture. This is where nuclei get bombarded by so many neutrons that they absorb them faster than they can decay, creating superheavy and neutron-rich elements like platinum and uranium. After the explosion, these newly minted elements are blasted into space.

Now, there's another fascinating place where the r-process occurs: the merger of two neutron stars. A neutron star is the tiny leftover core after a massive star explodes as a supernova. They tend to be one and a half times more massive than the Sun but only about fifteen miles across. When these dense stellar remnants collide, they eject incredibly neutron-rich material, which also undergoes rapid neutron capture. These mergers are now considered a significant source of the heaviest elements in the universe, especially after scientists observed one for the first time in 2017, an event that confirmed this hypothesis. That specific event produced an amount of gold equivalent to Earth's mass!

There's another lesser-known process called the p-process. This also occurs in supernovas, but instead of adding neutrons, it involves adding protons or using high-energy photons to eject neutrons from nuclei. This leads to the formation of some rare, proton-rich isotopes of elements like selenium and xenon.

Explosive stellar nucleosynthesis, as opposed to core stellar nucleosynthesis, happens in the final stages of a massive star's life after fusion in the core ceases. At this point, the envelope, previously supported by radiation pressure streaming out from the core, collapses onto the core, crushing it to a stiff ball of incredible density before rebounding. This rebound is the supernova explosion. The shock wave from the supernova blasts through the star's envelope, causing a rapid burst of fusion that creates many of the intermediate elements we see, like silicon, sulfur, and calcium. These elements are then spread throughout the universe. Every atom of gold, every molecule of oxygen, and every trace of carbon owes its existence to these cosmic events.

We are now getting closer to the middle of the Middle Realm, where we reside. But before we can get to us, we need a planet.

We need Earth. For us, this is the literal seat of the Realm of Life. We have only one home, but does that mean life can exist only on Earth?

Short answer: No. Far from it.

▪ ▪ ▪

So now we have our star, but we don't live on the Sun. We live on a planet made of stuff left over from the Sun's formation. The star-formation residue, if you will. Just as a star can be considered an inner core with a surrounding envelope, a protostar has a surrounding protoplanetary disk. Typically, the mass of the dense protostar is ten to a hundred times that of the disk, which may extend to a thousand times larger than the central protostar's radius. Eventually, winds from the forming star will blow away much of the disk. The remaining residual disk material flattens over time and forms planets.

Before diving into the details, I should define what I mean by "planet." My definition is much broader than that of the International Astronomical Union, because the drama of planetary

Fig. 7: Planetary Accretion
Dust clumps grow into planetesimals, which collide and merge to form protoplanets through a chaotic and violent accretion process.

science isn't limited to just the big-name "true" planets. Some of the most fascinating processes—volcanoes, oceans, atmospheres, magnetic fields, even the potential for life—are happening on moons and dwarf planets.

Suppose Galactus, the World Devourer, shows up in our solar system and decides to have Earth as a chilly dessert, so he places it in orbit around Jupiter to let it cool off. While he waits, he jaunts over to a neighboring solar system for his main course—the planet Proxima b. Would Earth cease to be a planet if it were now orbiting Jupiter instead of the Sun? I don't think so.

When I think of planet-ness, I think of a body sufficiently massive to have become gravitationally rounded into a sphere. The minimum mass required for a body to become rounded in this way depends on its composition. It must possess at least 0.01 percent (or one ten-thousandth) of Earth's mass if it is rocky; such a body would be about 200 miles across. If it is icy like Pluto (yes, by this definition, Pluto is considered a planet), it could be 70 percent less massive at 0.003 percent of Earth's mass, and would be about 150 miles across.

Our solar system contains 38 such bodies that we know of. These include the 8 major planets, 19 of their moons, 1 asteroid, and 10 Kuiper Belt objects (KBOs). We expect there to be around 120,000 icy bodies greater than sixty miles across in the Kuiper Belt—a region of icy bodies orbiting the Sun in a flat disk beyond Neptune's orbit. A significant number of these KBOs are expected to be gravitationally rounded. The Vera C. Rubin Observatory's Legacy Survey of Space and Time, which I worked on from 2008 to 2012 as a member of its early development team, and which began official operations in 2025, is expected to discover 40,000–100,000 new KBOs over ten years, significantly expanding our knowledge of this region.

Taken together, these rounded worlds—from Jupiter to much smaller KBOs and everything in between—are the ones I'll focus on for my discussion of the development of life as currently understood.

Let's look at Earth, the planet we know better than any.

Earth's journey began more than 4.5 billion years ago in the swirling chaos of our solar system's early days. At that time, the Sun, which was only about 100 million years old, was surrounded by a vast disk of gas and dust. Within this disk, tiny particles of dust were pulled together by electromagnetism, similar to how dust bunnies form under your bed. Once formed, the clumps continued to grow by colliding with each other, forming larger and larger clumps through a process known as accretion. The bigger clumps were larger targets and thus grew faster than the smaller clumps. They accumulated dust and gas from the surrounding material more quickly than the smaller ones. Eventually, these clumps grew large enough to become mini planets called planetesimals.

The largest planetesimals continued to grow, attracting more and more material until they became protoplanets. These protoplanets then entered a phase known as oligarchic growth. They were now large enough to increase their growth rate by gravitationally attracting even more matter. The largest began to dominate the surrounding region and attract the smaller ones into their orbit. Planetesimals collided and shared their materials. Some were incorporated, some were driven into the Sun, and some were ejected from the solar system altogether.

The "gas giants" of our solar system, Jupiter and Saturn, formed in this way. Even though they are primarily composed of hydrogen and helium, each possesses a core containing a mix of heavy elements (rock, metal, and hydrogen compounds) that formed through accretion. Once these cores were massive enough, their gravitational fields began to collect gas from the Sun's protoplanetary disk, and they got bigger and bigger.

Planet formation—including Earth's—is not a smooth pro-

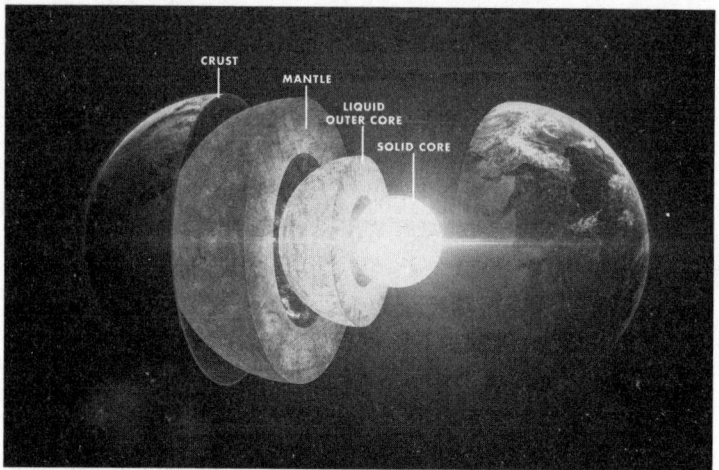

FIG. 8: EARTH INTERIOR
Earth's interior structure: a solid inner core, liquid outer core, viscous mantle, and solid crust—all critical to plate tectonics and habitability.

cess. Billions of years ago, our planet started as a hot and chaotic mess, likely containing material from neighboring protoplanets and planets. It was a molten ball heated from innumerable violent collisions. Over time, the molten rock differentiated into layers. Dense materials like iron, nickel, and iridium sank toward the planet's core, while less dense materials like silicon, carbon, nitrogen, and oxygen floated to the top.

But even as the surface of our planet began to cool and solidify, the interior remained incredibly hot. (Today, Earth's core temperature is roughly equivalent to the Sun's surface temperature!) As certain weighty, radioactive elements within Earth's core and mantle decay, they release tremendous heat, accounting for roughly half of Earth's heat-energy output. The other half is believed to come from residual heat within the core, which has been churning away for billions of years since Earth's formation. Together, these phenomena drive terrestrial processes such as plate tectonics and volcanic activity. Those activities, in turn, powered the emergence and evolution of life—and continue to do so today!

▪ ▪ ▪

There are three main things I'd ask you to remember as we journey beyond the Middle Realm to the other eight realms. First, electrons are foundational to our universe. Their existence is necessary for constructing matter; their behavior transports energy and creates light. Second, energy is the main currency of our universe. We've covered those.

Now we move to the third principle: Processes that occur spontaneously in our universe yield energy.

Energy becomes more concentrated in one place (like a star) precisely to make it less concentrated in another (like its surrounding protoplanetary disk). I like to say that the universe favors this "minimization of energy," or in science-speak, that systems evolve toward their lowest energy state. And that's accurate: When energy flows from higher to lower concentrations, the difference is what fuels change. This dynamic undergirds all creation.

At its most basic, the story of the Nine Realms—the story of the universe—is a story of energy flow. And here, in the Middle Realm, it's not just *whether* energy flows that matters but *how* it flows—and how concentrated or dispersed it is. That's where entropy comes in.

So, what is entropy? At its core, entropy measures how spread out energy is in a system. The more concentrated the energy, the lower the entropy. The more spread out and diffuse, the higher the entropy. A hot cup of coffee in a cold room cools down; that's intuitive. Energy spreads from the concentrated heat of the coffee to the surrounding cooler air. That's entropy increasing.

Now consider something less intuitive: the spontaneous formation of a star. That process seems to go against entropy, because energy is becoming more concentrated in the core. But here's the twist: While the energy in one region becomes more concentrated, the energy overall becomes more diffuse. The protostar heats up while radiating energy away, raising entropy in the surrounding region. That's the loophole. Local decreases in entropy are allowed so long as the total entropy of the system increases.

And sometimes, energy flows in directions that seem to outright defy our expectations.

Take heat flow. We're taught that heat always flows from hot to cold, spreading out to more entropy—and it usually does. That's the intuitive case. But can it spontaneously go the other way? According to virtually every physics textbook, no. But my experience—and recent research—suggests otherwise.

When I was a teenager, I discovered this while washing dishes. I'd dunk a hot skillet into cold water and hear it hiss and crackle as steam shot off. But then something strange would happen: The handle would suddenly get hotter, not cooler. I'd have to let go or risk getting burned. There was no external work involved—no heat pump, no electricity. Just metal, water, and a sudden jump in temperature at the wrong end of the pan.

Years later, a colleague of mine published a paper showing that spontaneous local reverse heat flow like this is, in fact, possible under the right conditions. These are the following:

1. Heat must be transferred from one substance to another— in this case, from metal to water.

2. The transfer must be large, meaning the temperature difference must be high enough to create strong conductive energy flow through the material.

Here's what's going on. Heat travels through metals primarily as vibrational waves. And when a wave hits the boundary between two different materials (like metal and water), some of it reflects. That reflected wave can carry energy backward, in this case toward the pan's handle, briefly heating it beyond its original temperature. No law is broken—just bent around a deeper principle.

This brings us to something crucial: Entropy isn't a dictator. It's a trend. It tells us where energy tends to go, not where it must go in every local moment. Systems can locally become more ordered, or energy can temporarily reverse direction—as long as the total entropy of the universe still increases. That's what makes life possible.

This is because, as with star creation, life is one of the great entropy paradoxes. It is the concentration of energy into complex, ordered systems that continue to persist, replicate, and evolve by riding energy gradients.

If you're wondering what keeps all this energy flowing—why the universe builds stars, planets, and molecules instead of just unraveling into chaos—the answer is this: It does both. Entropy doesn't prevent structure; it drives it. Creation happens not *in spite of* entropy but *through* it.

And nowhere is that paradox more vivid, more unlikely, or more beautiful than in what comes next.

The Realm of Life.

REALM OF LIFE

If evolution really works, how come mothers only have two hands?

—Milton Berle

Life is concentrated energy.

A single year of sunshine delivers more energy to Earth's surface than all of Earth's known fossil fuel reserves combined. Yet, for the time being, we primarily power our modern society using petroleum, coal, and gas.

Fossil fuels possess remarkably high energy concentrations. Take a gallon of gasoline. This meager volume of liquid possesses the capacity to propel a vehicle weighing over two tons a distance greater than twenty miles at a considerable speed.

Fossil fuels, such as gasoline, derive their name from their origin as the remnants of ancient life, including wetland forests transformed into coal and algae and zooplankton transformed into petroleum. It is truly striking to consider that the force driving much of our modern society is rooted in the energy stored in past life-forms.

But why does life exist at all? Unsurprisingly, I am not the first scientist to ponder this question or seriously consider its answer. There is a theory that explains the emergence of life as a natural consequence of molecules self-organizing to absorb and then dissipate heat energy. In other words, life spontaneously comes into existence specifically to facilitate the process of energy leaving matter—the idea of minimization mentioned at the conclusion of the previous chapter.

The origin of life on Earth remains a topic of much scientific debate and speculation. But there are some elements of the process that we have deciphered. What is certain is that Earth's geological conditions in its early history constrained how, when, and where life began on this, our home. Once formed, life became a geological force, modifying the planet and altering itself in response, creating a never-ending process of change, adaptation, and the emergence of greater and greater biological complexity. (You might think I'm talking about us here, with all the recent and well-founded alarm over climate change and fancy words like "Anthropocene" coming into the lexicon, but really, I'm talking about any form of life that can modify the planet's chemistry. For example, cyanobacteria perform photosynthesis, modifying atmospheric and ocean chemistry—and they ain't us.)

But we're getting ahead of ourselves. Before organic evolution, there was molecular evolution. As we have learned, stars synthesize most elements. Some elements are created in the process of core fusion. Some are made in envelopes of old stars. Some are created when stars explode. And some are created when the cores of dead binary stars collide—all the gold you're wearing, for example, comes from this process. However, none of these processes produces the complex molecules on which life depends.

Nature excels at forming simple molecules consisting of one or two types of atoms. You are already familiar with many of these abundant, small molecules. Nitrogen (N_2), oxygen (O_2), and hydrogen (H_2) gases are common molecules consisting of only one type of atom. Substances like water (H_2O), carbon dioxide (CO_2), methane (CH_4), and ammonia (NH_3) are common molecules consisting of two types of atoms.

The types of molecules life uses, such as amino acids for

building proteins and nucleotide bases for RNA and DNA, contain many more atoms and are much more complex. Due to their complexity, they are harder to build, easier to destroy, and much rarer than nature's abundant, simple molecules. For life to take hold, there must be an evolutionary process that uses simple molecules as the starting material and, over time, builds more complex molecules.

This type of molecular evolution proceeds by adding energy to small molecules, breaking them apart, and then allowing them to recombine spontaneously into more complex forms until a self-replicating molecule is formed. In a famous set of experiments, scientists in the 1950s filled a flask with water, methane, and hydrogen and then added energy through an electrical spark. They didn't produce self-replicating molecules, but they did produce eleven of the twenty amino acids on which Earth life is based.

In the decades since the 1950s, origin-of-life experiments have become considerably more sophisticated. In 2018, researchers in Ontario, Canada, created a machine called the Planet Simulator, which mimicked conditions on early Earth. After running the experiment for only a couple of months, the researchers discovered that it had spontaneously produced "protocells" and RNA strands. Their work confirmed the basic idea behind the RNA world hypothesis, a leading theory for the development of early life. They concluded that life is "probably a relatively frequent process in the universe."

If this is true, what are the minimum requirements for basic life? And what constitutes basic life?

First, we need a self-contained vessel to form a cell membrane, which creates an enclosed volume separate from the rest of existence. A cell's membrane is what allows life to stay organized and become more complex, even though nature tends to favor disorder. The second law of thermodynamics says that, over time, things naturally break down and become more random. But inside a living cell, the opposite happens: The cell builds structures, carries out precise chemical reactions, and maintains order. This isn't magic; it results from the membrane's ability to control what happens inside.

The membrane surrounds the cell's internal environment, creating a space where it can organize itself without interference from the outside. It acts like a manager, ensuring that the right molecules are directed to their intended locations while maintaining balance between different cellular components. Inside, the cell is filled with tiny compartments, each with its own function, and the membrane helps control the movement of energy and materials between them. This allows the cell to store energy, build complex molecules, and conduct reactions in a controlled manner, rather than allowing everything to mix randomly.

Because of this controlled organization, the cell isn't overwhelmed by entropy. Instead, it channels energy into creating order, keeping itself running smoothly, and even evolving into more sophisticated forms over time.

For Earth life, cell membranes are formed from a type of fat molecule known as a lipid. Lipids form abundantly in nature and have a remarkable property: When placed in water, they spontaneously self-organize into spherical structures—essentially primitive membranes. These kinds of lipid-based protocells are exactly what the Planet Simulator experiment produced. The fact that they formed so readily in a simulated early Earth environment suggests that protocells may not be rare or difficult to generate. On the contrary, they might be one of the easiest and most natural steps toward life.

Speaking of water, a liquid solvent is needed because it provides a flexible, dynamic environment where molecules can move, interact, and undergo the chemical reactions necessary for life. A solid would lock molecules in place, preventing them from mixing and reacting, while a gas would be too chaotic, with molecules moving too fast and too far apart to sustain stable interactions. (And you can forget about plasma—it's way too crazy in there!) A liquid, however, strikes the right balance: It allows molecules to dissolve, come together, and react in a controlled yet adaptable way.

Water, the liquid solvent of life on Earth, is particularly good at this because it can dissolve a wide range of substances, making it easy for cells to transport nutrients, remove waste, and maintain a stable internal chemistry. It also helps proteins and other biological molecules fold into the right shapes to function properly, and it supports the flow of energy that cells need to survive. Water is also abundant not only on Earth but throughout our solar system and the universe.

Hypothetically, other liquids could support life in different conditions. For example, some scientists speculate that methane, which is liquid at extremely cold temperatures, might serve as a solvent for alien life. Laboratory experiments have shown that many of the molecules utilized by life are stable in more exotic liquids—including ammonia and even sulfuric acid (the *Alien* alien, with acid for blood, could be real!)—so water isn't strictly necessary, but a liquid is. Life needs a medium where molecules can move and interact in a way that allows complexity to emerge and sustain itself.

After a vessel and a liquid, the next two ingredients required are large, complex genetic messenger molecules like RNA and DNA, and proteins necessary for structure and metabolism.

The Planet Simulator experiment showed that these molecules can form spontaneously in nature, and not just on Earth. As of 2022, all five nucleotide bases that form RNA and DNA have been found in meteorites. As for proteins, we've known since the 1950s that the amino acids used by proteins can be spontaneously produced.

Other experiments provide even more compelling data. In 2022, researchers analyzing samples from the asteroid Ryugu discovered more than twenty types of amino acids. This finding was particularly exciting because the Hayabusa2 probe collected pristine *subsurface* asteroid material. Before landing on the asteroid, Hayabusa2 deployed a small explosive device called the small carry-on impactor. This device launched a copper projectile into Ryugu's surface at high speed, creating a crater and exposing fresh subsurface material.

Following this operation, the spacecraft descended to the newly formed crater area. Hayabusa2 then used its sampling horn to collect the excavated subsurface material. During this process, a five-gram metal bullet was fired into the surface, stirring up debris that was funneled into the spacecraft's sample container. This method ensured that pristine material from beneath Ryugu's weathered outer layer was captured for analysis back on Earth, untainted by Earth's atmosphere or microorganisms, providing a pure glimpse into the asteroid's composition.

In early 2025, scientists made a remarkable discovery in the samples returned by NASA's OSIRIS-REx mission from the asteroid Bennu. They identified thirty-three different amino acids, including fourteen of the twenty amino acids life uses to build proteins. This finding is extraordinary because it represents the most diverse array of amino acids ever found in extraterrestrial material. The Bennu samples were also found to be rich in carbon, nitrogen, and ammonia, suggesting that asteroids might have played a crucial role in supplying the raw materials for life on our planet.

And now the search is tightening closer to home. In September 2025, NASA announced what it called "the clearest sign yet" of possible life on Mars, based on chemical fingerprints preserved in the ancient mudstones of Jezero Crater. The Perseverance rover discovered that organics and reactive minerals, such as iron and sulfides, are bound together in ways that, on Earth, typically trace back to microbial activity. While non-biological explanations remain on the table, the discovery raises the possibility that simple life not only arose on our neighboring planet but left behind signatures we can still read billions of years later. At the moment, this is just circumstantial evidence, similar to circumstantial evidence we've found on our planet that hints at Earth's earliest life. But in the case of the Martian samples, the circumstantial evidence is stronger, and it could be conclusive if we can safely return the samples to Earth for detailed study.

These findings don't just tell us about the past; they also hint at the potential for life elsewhere in the universe. At a minimum, the universe appears to be brimming with protocells, amino acids, and RNA.

The process of transforming nonliving matter into self-replicating cells is one thing. Going from these cells to an advanced technological civilization like ours is another matter entirely. And here is where Earth's rare conditions come into play.

To consider how humans came to exist, we must again indulge the absurd and escape some of our normal deception. Virtually every single animal on Earth and most fungi require oxygen for aerobic respiration. This is a metabolic process that takes glucose and oxygen and converts them into water, carbon dioxide, and energy in the form of ATP molecule chemical bonds (these bonds provide energy for other cellular functions, such as those that help our muscles contract, our nervous systems transmit signals, and our cells make other cells).

Likewise, all nonparasitic plants and algae require sunshine for photosynthesis, which happens to run counter to respiration: Here, plant cells take water, carbon dioxide, and energy (in the form of sunlight) and convert these into glucose and oxygen. These two complementary processes are vital to the vast majority of life on Earth.

But when life on Earth was just getting started, neither oxygenic photosynthesis nor aerobic respiration had developed. Those early cells were simple and rudimentary, and they couldn't deal with oxygen, which is highly reactive (just think of what oxygen does to iron). In fact, oxygen was lethal to early life-forms on Earth. Complex Earth life, which requires oxygen and which produces it in abundance, had to evolve. And as luck would have it, it did evolve right here on planet Earth.

Why? You might think it's because of our abundant water. But that's not enough. Lots of bodies in our solar system contain water. At least two moons of Jupiter—Ganymede and Europa—have even *more* water than Earth.

What makes Earth special is that our liquid water is bathed in sunlight.

So far, we have identified nine other ocean worlds in our solar system: Europa, Ganymede, Callisto, Titan, Enceladus, Pluto, Triton, Mimas, and Ceres. The oceans of these worlds are covered by miles of ice, rock, or a completely opaque atmosphere—and sometimes all three. Only one of these boasts abundant surface liquids of any sort: the Saturnian moon Titan. But Titan's atmosphere is so thick that 90 percent of the light that reaches the tops of its clouds is absorbed before making it to the surface. Titan is ten times farther from the Sun than Earth, which means that the tops of Titan's clouds receive only 1 percent of the sunlight that Earth does, which in turn means that Titan's surface receives only 0.1 percent of Earth's surface irradiance. It's dark down there! If indigenous conscious beings are walking (or rolling or swimming or flying) around on Titan, they don't even know that stars exist!

These conditions are typical not only of our planetary system but of most others as well. They are predominant. In recent years, we have discovered approximately six thousand planets around other stars, plus the thirty-eight spherical bodies within our solar system. This sample shows that atmospheres usually occur in one of two basic configurations: superthick or virtually absent. Earth is the outlier.

Earth has a transparent atmosphere that is sufficiently thin to allow sunlight to penetrate to its surface. This means that once life began on Earth and reached near the water's surface, it was bathed in sunlight. The first photosynthetic life-forms on Earth likely evolved around 3.4 billion years ago, during the Archean Eon.

However, these early photosynthesizers weren't the oxygen-generating kind. They used anoxygenic photosynthesis, relying on molecules like hydrogen sulfide (H_2S) instead of water, and they didn't release oxygen. These are your sulfur-loving purple and green bacteria — the OG solar panel gang, but with zero O_2 emissions.

Oxygenic photosynthesis, the type that splits water and releases oxygen (thanks, cyanobacteria), evolved a bit later— somewhere between 2.7 to 3 billion years ago. They began releasing oxygen as a by-product of their metabolism. But oxygen did not immediately flood the atmosphere. Instead, it reacted with vast amounts of iron and other ocean minerals, vanishing as quickly as it was produced. For hundreds of millions of years, oxygen remained locked away in these chemical reactions, forming thick layers of iron-rich rock on the ocean floor.

It wasn't until approximately 2.5 billion years ago that the balance shifted, marking the beginning of the Great Oxidation Event (GOE), which led to the first significant accumulation of oxygen in the atmosphere. However, this wasn't a rapid or smooth rise; it was a slow, stuttering process that unfolded over at least 200 million years. Oxygen levels surged and retreated as Earth's chemistry adjusted to the new reality. Even after the GOE, oxygen remained scarce in the deep ocean for nearly 2 billion more years. It wasn't until 600 to 800 million years ago that oxygen finally reached levels high enough to support the first complex multicellular life.

With the rise of oxygen, another critical process began—the formation of the ozone layer. In the upper atmosphere, ultraviolet (UV) radiation from the Sun splits oxygen molecules into individual atoms, which are then recombined with O_2 to form ozone (O_3). Before this, Earth was bombarded by intense UV radiation, making our planet's hard, non-liquid surface inhospitable for life. As the ozone layer thickened, it absorbed much of this harmful radiation, creating a protective shield that would eventually allow life to move beyond the oceans and colonize dry land.

The story of oxygen on Earth is about not just biology but the profound, intricate interplay between life, the planet, and time itself. The microbes that first released oxygen set off a chain reaction that would reshape the world, but the transformation was anything but quick. It took nearly two billion years of slow accumulation, geological shifts, and biological adaptation before oxygen could finally dominate the atmosphere and create a world capable of supporting the future explosion of life. And perhaps the most underappreciated player in this interaction is Earth's invisible force field—its magnetosphere.

The magnetosphere is the name we give to the protective magnetic bubble that surrounds Earth and protects our thin and transparent atmosphere from erosion. It exists because Earth's rotation and internal heat flow generate a constant movement of hot molten iron and nickel in its outer core. Earth's solid inner core stabilizes and magnifies this field, forming a self-sustaining system.

The wind that streams from the Sun is constant, subjecting planetary atmospheres to a never-ending bombardment of invisible charged particles, primarily protons and electrons. When these particles encounter a planet, they can interact with its atmosphere. Over long periods, this interaction gradually wears away or erodes the atmosphere, especially on planets that lack a robust magnetic field.

But Earth's magnetosphere deflects most of the charged particles, like water flowing around a boulder in the middle of a small stream. If not for this phenomenon, solar wind and cosmic radiation would have stripped away Earth's atmosphere long ago, leaving its surface dry, barren, and lifeless, like on Mars.

Scientists have compelling evidence that Mars once possessed a much denser atmosphere than it does today. This conclusion is supported by numerous geological features on the planet's surface that indicate the presence of flowing water in the distant past, including ancient riverbeds, lake beds, and what appear to be remnants of vast seas. However, Mars lacks a robust magnetic field. Over time, persistent bombardment by the solar wind has gradually stripped away much of Mars's original atmosphere, leaving it significantly thinner than it once was. NASA's MAVEN (Mars Atmosphere and Volatile Evolution) mission, launched to study the Martian atmosphere and its interaction with the solar wind, has provided concrete evidence that this process continues to this day.

Venus presents a striking contrast to Mars in terms of atmosphere. While Mars has a thin, fragile layer of air that has gradually been stripped away, Venus is enveloped by an incredibly dense atmosphere, composed primarily of carbon dioxide, a greenhouse gas that traps heat with extreme efficiency. This thick atmosphere, which is around ninety-three times more massive than Earth's, has been largely intact for billions of years, making Venus far less vulnerable to the kind of atmospheric erosion that dramatically altered Mars.

Unlike Earth, which generates a strong magnetic field through the movement of molten iron and nickel in its core, Venus lacks such an internal dynamo. Instead, it has what is known as an induced magnetic field, formed through the interaction between its dense atmosphere and the solar wind. As the wind particles collide with Venus's upper atmosphere, they strip away electrons, creating a region of charged plasma. This plasma, in turn, generates weak magnetic fields that partially deflect the solar wind, forming a thin, elongated magnetic tail behind the planet. While this induced field offers some degree of protection, it is far less effective than Earth's internally generated magnetic shield at preventing atmospheric loss.

Despite its weak magnetic defenses, Venus's thick atmosphere has proven remarkably resistant to being stripped away. Its sheer density is a powerful shield, preserving much of the planet's air over time. But this atmospheric heaviness comes at a steep cost. The overwhelming presence of carbon dioxide has fueled an extreme greenhouse effect, pushing surface temperatures to a staggering 864°F (462°C)—hot enough to melt lead. This makes Venus's surface an unrelenting inferno, vastly different from both Earth's habitable conditions and certain cold but potentially life-supporting regions of Mars.

Venus serves as a stark example of how a planet's atmosphere can influence its destiny. Similar in size, mass, and composition to Earth, young Venus had the potential to support abundant life. However, it followed a very different path. Its dense atmosphere has prevented complete atmospheric loss, but this same atmosphere, rich in greenhouse gases, has turned the planet into one of the most extreme environments in our solar system. The sweltering conditions on Venus highlight the dangers of runaway greenhouse effects, demonstrating how a once potentially habitable world can become an uninhabitable wasteland.

So again, why does Earth have a strong magnetic field while Venus and Mars do not? The simple answer: a big-ass astronomical collision. The universe smacked Earth into a state of creepy fertility.

I don't mean "weird" creepy. I mean "gradual movement" creepy. Below our feet, Earth's solid silicate mantle—like the envelope to Earth's core—slowly creeps up in some places and down in others. This process occurs due to temperature differences within the mantle, causing hotter, less dense material to rise, and cooler, denser material to sink. At Earth's surface, this flow manifests as plate tectonics. This movement of hot material from beneath transfers heat from Earth's interior to its surface, ultimately radiating into space as infrared light.

Before we move forward, let's take a minute to truly visualize planet Earth. We think of Earth as solid to the core, but it is not. The solid, flowing mantle constitutes the outermost 45 percent of Earth's radius. The solid core takes up the innermost 19 percent of Earth's radius. Between the two, accounting for 36 percent of Earth's radius, is the liquid outer core, consisting entirely of molten metal.

By volume, there is 30 percent *more* liquid metal inside Earth than all of the surface water combined!

The temperature difference can be hundreds or thousands of degrees where the bottom of the solid mantle meets the top of the liquid outer core deep inside Earth's interior. That difference in temperature sets up a powerful heat drain. The outer core loses energy to the mantle. This heat loss stirs up motion in the outer core. The liquid iron-nickel mix moves because hot material rises, cools off near the boundary, and then sinks again, driving a constant churning motion. As the molten metal flows, it generates electric currents, creating our self-sustaining magnetosphere through a process called the geodynamo.

How did these convective motions get started? That big-ass cosmic smack I mentioned is how! Approximately 4.5 billion years ago, during the chaotic infancy of the solar system, a Mars-sized protoplanet named Theia collided with proto-Earth.

Computer simulations suggest that the collision likely occurred at a moderate velocity and an oblique angle. Theia's iron core merged with Earth's core, while much of its mantle material combined with Earth's mantle. Simultaneously, a significant amount of debris was ejected into orbit around Earth, eventually coalescing to form the Moon. This process explains why lunar rocks share striking similarities with Earth, indicating a thorough mixing of materials from both celestial bodies during the impact.

The immense energy from the collision stirred Earth's molten core, initiating the vigorous convection that powers its geodynamo. Additionally, the energy delivered by the impact likely prevented Earth's core from becoming stratified or stagnant, ensuring its dynamic behavior over billions of years—a behavior that continues to this day.

Even with its magnetic shield, Earth's atmosphere is constantly in flux. Atmospheric gases escape into space, while new ones are added through geological, mainly volcanic, activity. Gases also seep out from mid-ocean ridges, hot springs, and deep underground reservoirs. This ongoing exchange has kept our planet's air relatively stable for billions of years, with volcanic emissions generally outpacing the loss of particles to space.

The upshot: Earth's thin, transparent atmosphere, protective ozone layer, and strong magnetic field together form a three-layered filter—one that screens out the deadly and lets the life-giving through. It blocks the Sun's harshest radiation while allowing visible light to flood in, warming our oceans without boiling them away. That light, filtered and softened, reaches the surface and interacts with liquid water, the most chemically fertile solvent in the universe. This is the setup. Sunlight bathing stable surface water, shielded just enough to last. That's the recipe for complex, multicellular life. This is the miracle. And as far as we know, it happened only here.

▪ ▪ ▪

Earth is special. Evidence suggests that Earth has boasted some form of life for a long, long time. The earliest evidence of life on Earth comes from a remarkable discovery involving carbon isotopic ratios found inside ancient zircon crystals. Zircons are incredibly resilient minerals, and some found in Western Australia's Jack Hills date back nearly 4.4 billion years, making them among the oldest pieces of Earth's crust. Within these zircons, tiny inclusions of ancient material hold clues about Earth's early environment, including possible signs of life.

You've probably heard the term "carbon dating" before, but it's not about carbon going out to dinner and hooking up; it's about figuring out how old stuff is. Analyzing carbon isotopes is the key to detecting signs of early life in these ancient zircons.

Here's how it works. Carbon exists in two stable isotopes: carbon 12 (^{12}C) and carbon 13 (^{13}C). Living organisms prefer the lighter isotope, ^{12}C, creating a distinct "signature" in the ratio of these isotopes when searching for past signs of life.

In this case, scientists found tiny inclusions of graphite (a form of carbon) inside zircons dating back to about 4.1 billion years ago. When they measured the carbon isotopic ratios in these inclusions, they found a pattern that suggested the presence of life, with ratios similar to those seen in biological carbon on Earth today.

Don't forget: Earth is 4.5 billion years old, meaning this evidence suggests it took less than 400 million years for basic life to get living! Earth was no picnic back then; in fact, young Earth was quite hostile to life. Regardless, something—something quite small and prevalent—appears to have been living at that time. This discovery suggests that life could have emerged relatively quickly after Earth's surface cooled enough to support liquid water, around the time of the Late Heavy Bombardment, a period when the planet was still being pummeled by asteroids and comets swarming our young solar system.

It's important to note that the carbon dating evidence from these zircons does not show direct fossils or organisms, so we don't know what these early life-forms looked like, but it does point to strong evidence of biological processes that likely involved early microbial life.

And as you might have guessed, the action was not limited to zircon crystals found in modern-day Australia. In Greenland, scientists have discovered rocks dating back to approximately 3.8 billion years ago that exhibit carbon isotopic signatures indicative of biological activity. These ancient rocks also contain graphite enriched in ^{12}C. Like the zircon data, these are not direct fossils but only circumstantial evidence of early life.

The oldest direct examples of evidence for life come from stromatolites, which are layered, rocklike structures built by communities of microorganisms, particularly cyanobacteria (also known as blue-green algae). These microbes trap sediment and build up mounds or mats over time, forming distinctive rock patterns.

Fossilized stromatolites found in Western Australia date back to approximately 3.5 billion years ago. These structures indicate that early microbial communities thrived in shallow, sunlit waters at least this far back in Earth's history. Modern stromatolites still exist in places like Shark Bay, Australia, serving as analogues for these ancient ecosystems.

What brings together the circumstantial isotopic data and the direct evidence from fossilized stromatolites are recent discoveries in the field of genetics. In 2024, by analyzing the genomes of hundreds of bacteria and archaea, researchers identified the last universal common ancestor (LUCA) to all life on Earth, which existed approximately 4.2 billion years ago. This LUCA was anaerobic, meaning it lived without oxygen—an adaptation that makes sense, since Earth's atmosphere and oceans at the time lacked free oxygen entirely—but it was still advanced enough to possess an immune system, suggesting it had to protect itself from other rudimentary life-forms and viruses. LUCA was not alone!

Once life emerged on Earth, its evolution was far from smooth. It was shaped by a chaotic interplay of geological upheavals and environmental extremes, with each challenge testing its resilience. It would take *another* two billion years for the first oxygen-producing photosynthesizing organisms to evolve.

During the Late Heavy Bombardment, approximately four billion years ago, Earth was struck repeatedly by asteroids and comets, some of which carved massive craters and unleashed energy on unimaginable scales. Oceans vaporized, the crust melted, yet microbial life remarkably endured. It might have survived deep underground, within hydrothermal vents, or in other protected niches, lying dormant like seeds buried in ash until conditions improved. This resilience suggests that life had already developed the evolutionary capacity to recover and persist even in the face of near-total annihilation.

As Earth stabilized, it swung to another extreme: the Snowball Earth episodes between 2.4 and 2.2 billion years ago. During these periods, entire oceans froze over, glaciers reached the equator, and the planet became an icy tomb. Yet life persisted, clinging to warmth near volcanic activity beneath thick ice where liquid water remained insulated, or in deep-sea hydrothermal vents. When the glaciers eventually retreated, they released a flood of nutrients into the oceans, triggering a series of evolutionary booms. These global freezes acted as nature's reset buttons—harsh transformative events that reshaped the trajectory of life.

Following these cataclysms was the so-called Boring Billion, a period spanning roughly 1.8 to 0.8 billion years ago. Although it lacked dramatic environmental upheavals, this era was far from uneventful. Oxygen levels remained low but fluctuated enough to create selective pressures that encouraged the evolution of complex cells—eukaryotes. Evidence indicates that multicellular life emerged multiple times during this period but often failed to persist. Fossil and genetic data reveal early experiments in multicellularity that were ultimately unsuccessful—prototypes abandoned in an evolutionary workshop. These organisms likely faced challenges such as energy shortages or environmental instability, making long-term survival unsustainable.

The repeated rise and fall of multicellular life during this era highlights a profound truth: Multicellularity is not a rare accident but an iterative process. Evolution repeatedly tests it, and even today echoes of these ancient trials can be seen in organisms like slime molds and colonial bacteria that blur the line between single-celled and multicellular life. For complexity to take hold permanently, however, specific conditions must align—just as they finally did before the Cambrian explosion when multicellular organisms diversified into the extraordinary array of life-forms we see today.

But evolution didn't stop at cells joining forces. It continued to stack complexity: building neural networks and eventually creating brains. And somewhere along the way, those brains stopped only reacting to the environment and started simulating it. Modeling it. Anticipating it. That shift, from adaptation to imagination, is where biology gives rise to minds. And not just any minds: ones that can reflect, predict, and create. That's a different kind of leap. A realm all its own—the Realm of Imagination.

Through bombardments, deep freezes, and long stretches of seeming stagnation, Earth life has persisted against all odds. Its evolution hasn't been a steady march forward but more like a braided river—branching, looping, sometimes vanishing before surging ahead again. Life's history isn't just survival; it's relentless experimentation, where failure sets the stage for future success.

Fortunately, life on Earth began with the most adaptable units imaginable: prokaryotes and viruses. Prokaryotes—bacteria and archaea—are single-celled organisms without a nucleus. Their simplicity makes them incredibly efficient and resilient. They've dominated nearly every environment on Earth for billions of years. While bacteria and archaea look similar under a microscope, their genetics tell us they're profoundly different.

Viruses, though technically not alive, have played a crucial role in life's evolution as well. They introduce genetic variation, mutate rapidly, and exchange genetic material through a process known as recombination. Some, like retroviruses, go even further: They insert their genetic code into their host's DNA. Over time, these viral stowaways became permanent residents. Approximately 8 percent of the human genome is of viral origin—far more than we inherited from the Neanderthals and Denisovans with whom we interbred. Some of these ancient viral genes were repurposed by evolution, influencing brain development and immunity and even enabling live birth in mammals.

Bacteria and archaea also evolved stunning survival tactics. Some form spores and wait out harsh conditions—sometimes for millions of years—before regenerating. Researchers have even found what may be living cells in 830-million-year-old rock. These ancient microorganisms—viral and cellular—are the roots of the evolutionary tree from which all life descends. You're part bacteria, son. (Really. By cell count, you're mostly microbial. You're literally buggin'.)

Life on Earth includes six kingdoms plus viruses. Two kingdoms—bacteria and archaea—are prokaryotic. The other four—plants, animals, fungi, and protists—are eukaryotic. Eukaryotes evolved a key upgrade: compartmentalization, a.k.a. the development of membrane-bounded internal cellular compartments. Eukaryotic cells have nuclei to manage DNA and specialized organelles for various functions. A major evolutionary leap occurred when an early archaean engulfed an oxygen-using bacterium but didn't digest it. This partnership created mitochondria, greatly boosting cellular energy. This event, eukaryogenesis, set the stage for the development of complex life.

Why was this necessary? Because prokaryotes faced a fundamental constraint in developing complexity. For two billion years, they could increase complexity simply by growing larger; however, this approach eventually reached a critical limit. The problem lies in how prokaryotic cells generate energy: They produce it at their surfaces. As a cell grows, its volume expands faster than its surface area, which means internal energy demands increase more rapidly than the cell's capacity to generate energy externally. This creates an energy deficit where larger cells cannot produce sufficient energy at their limited surface area to power their expanding internal activities.

For nearly two billion years, this energy bottleneck limited life's complexity. Prokaryotes flourished but remained small and structurally simple. Life was essentially stagnant at this level until the arrival of mitochondria. Mitochondria solved the energy problem by generating energy inside cells, greatly expanding internal energy-producing surfaces. This breakthrough enabled cells to grow larger, support larger genomes, develop specialized compartments and internal structures, and ultimately collaborate to form multicellular organisms.

This energy revolution unlocked unprecedented biological complexity. Later, a similar symbiotic event led to the formation of chloroplasts, which are the engines of photosynthesis in plants and algae, when eukaryotes engulfed cyanobacteria. Like mitochondria, chloroplasts retain traces of their bacterial ancestry.

Another mystery is the origin of the nucleus. One idea is that the cell's membrane folded inward to form a new compartment. Another proposes that a large DNA virus infected an archaeal cell and remained permanently, evolving into the nucleus. Either scenario demonstrates evolution's creativity in remixing genetic and structural materials.

Some eukaryotes went beyond solitary existence, forming multicellular organisms. Early multicellularity was simple, likely resembling slime molds or bacterial colonies. Complex multicellular life emerged much later, diversifying significantly during the Ediacaran Period (635–539 million years ago), followed by the Cambrian explosion (539–487 million years ago), when ancestors of modern animal groups rapidly appeared.

Multicellularity evolved independently multiple times. Plants, animals, fungi, and algae separately achieved multicellular complexity through convergent evolution. Once the energetic barriers vanished, multicellularity offered advantages repeatedly discovered by different groups.

Despite this evolutionary breakthrough, widespread multicellularity took hold only about 635 million years ago. Consider that Earth formed 4.5 billion years ago, hosted oceans by 4.4 billion years ago, supported simple life more than 4.2 billion years ago, and saw eukaryotes arise 2 billion years ago. Despite ideal conditions and numerous adaptive life-forms, Earth took an astounding 2.5 billion additional years to evolve multicellular organisms.

The key takeaway is this: Simple life is resilient and probably common throughout the cosmos, but complex life requires substantial time, abundant energy, and an extraordinary dose of cosmic luck.

■ ■ ■

How widespread do I expect multicellular life to be in our galaxy? To calculate this, we can use the lessons we've learned about how multicellular life formed on Earth to create a modified Drake equation.

The original Drake equation, a formula used to estimate the number of technologically advanced civilizations in the Milky Way, was proposed by the astrophysicist Frank Drake in 1961. It consists of seven terms. Each term represents a factor involved in the emergence of intelligent, technologically communicative life. While the equation does not provide a definitive answer, it frames the problem in a structured manner, transforming a philosophical question into a scientific one. Although we cannot assign precise values to many of the terms, the equation remains a foundational tool in astrobiology and the search for extraterrestrial life.

It looks like this:

$$N = R_* \cdot f_p \cdot n_e \cdot f_l \cdot f_i \cdot f_c \cdot L$$

Before we dive in, don't worry! I promised not to overwhelm you with math on our journey, and I'm not reneging on that. For my math-phobes, know up front that this is not a quantitative equation so much as a qualitative one (ditto for all the other equations found in the rest of this chapter).

So, what are its qualities? The term we're solving for, N, is the number of intelligent, technologically advanced civilizations in the galaxy. The terms that lead us to N are the rate of star formation (R_*); the fraction of stars with planets (f_p); the number of planets that can support life (n_e); the fraction of those that develop life (f_l); the fraction of life worlds with intelligence (f_i); the fraction of those that develop detectable technologies (f_c); and the length of time these technologies are detectable (L).

Below, I propose a similar equation that differs philosophically. Drake based his equation on outcomes, many of which he knew precious little about. Since the 1960s, when Drake was active, we've learned a great deal about phenomena such as the rate of star formation, the prevalence of planets, and the types of environments that are most likely to support basic life. I will modify and update the equation, basing it not on outcomes but on probabilities of the galactic, stellar, and geological drivers for life's development to be present in a stellar system.

My equation differs from Drake's in three ways. As mentioned above, the first is that it is based on conditions rather than outcomes. The second—and this is related—is that I'm writing my equation more than six decades after Drake. I've seen modern results in biology, astrobiology, and planetary science. I can make inferences that were not so obvious in the 1960s. For example, Drake has a term for the fraction of stars that host planets. We now know that nearly every star hosts planets. The third is that, rather than looking for intelligent, technologically advanced civilizations, my equation solves for planets that can host complex, multicellular life.

First, we need a star in the right part of any given galaxy. Our galaxy contains a four-million-solar-mass black hole at its center, spewing out radiation and wreaking havoc. A star must be suffi-ciently far away from that monster to host planets that are safe for life.

Next, we need a star containing sufficient heavy elements (which as we know come from dead stars) in its protoplanetary disk to build a rocky planet containing the ingredients for life. Our Sun, for example, comprises material from around three dozen dead stars. We also need the six essential elements of life: nitrogen (N), carbon (C), hydrogen (H), oxygen (O), phospho-rus (P), and sulfur (S)—N-CHOPS.

Next, we need our star to be the right size. Massive stars burn fast and die young; the most massive stars last less than ten mil-lion years, which, as we've just learned, isn't close to enough time for complex life to form. On the other hand, small stars live a long time, but their light output is varied, tumultuous, and fe-vered. Their frequent eruptions of highly energetic radiation and matter would destroy life on a nearby planet, regardless of the strength of any planet's magnetic field. So we need a star that's big enough to be a source of stable light but small enough to last for several billion years.

Next, we need a planet with liquids, radiation shielding, a source of geological energy to get life started, and an atmosphere thin enough to let light through so that life can evolve oxygen-producing photosynthesis.

Considering these factors along with the rate of star forma-tion, we can construct the following five-term equation for the number of stellar systems, N_{mcl}, that bear multicellular life.

$$N_{mcl} = N_* \cdot S \cdot P \cdot f_l \cdot \gamma$$

Here we have the number of stars in the galaxy (N_*); followed by the fraction of "just right" stars (S) and "just right" planets (P); next is the fraction of worlds that will develop life (f_l); and finally we have the fraction of these worlds that have the right light, magnetic field, and chemistry to develop multicellular life (γ).

Three of these terms—S, P, γ—have their own set of terms. Let's start with S, the number of stars suitable for life in any given galaxy.

$$S = f_g \cdot f_Z \cdot f_P \cdot f_{\Delta t} \cdot f_{\Delta \lambda}$$

The first term is the fraction of stars far enough away from every galaxy's supermassive black hole for life even to be possible (f_g); then we have the fraction of stars containing sufficient heavy elements to build rocky planets (f_Z); then we have the fraction of stars containing phosphorous (f_P), the rarest of the six essential elements of life; followed by the fraction of stars with a sufficiently long lifetime for life to develop and evolve ($f_{\Delta t}$); and finally, we have the fraction of stars whose light output is stable and useful ($f_{\Delta \lambda}$).

Now let's turn our attention to the fraction of just-right planets, P. Traditionally, when astro-folks communicate life-friendly planetary conditions, we invoke the habitable zone (HZ) concept. The HZ is a region surrounding a star where bodies of liquid water could exist on a planet's surface, given that the planet has matching atmospheric conditions. While that may be the ideal condition for developing oxygen-producing photosynthetic life, these conditions are unnecessary to develop simpler prokaryotes and viruses. The more fundamental requirement is that liquid reservoirs exist. If so, life has a chance, even if all that liquid dries up or freezes occasionally.

This all means that the conditions for developing life are likely more widespread than we have typically accounted for by focusing on the habitable zone, since objects outside can still provide habitable niches. Considering this, I present a three-term equation for the drivers determining the fraction of planets, P, capable of supporting life:

$$P = f_{lq} \cdot f_{r-} \cdot f_{geo}$$

The first term is the fraction of planets possessing liquids (f_{lq}); the second is the fraction of planets with radiation shielding (f_{r-}); and the third is the fraction of planets that meet basic geological conditions (f_{geo})—as far as we currently understand them—to support the creation and evolution of simple life.

The final of these three terms, γ, is the luck term and can be expressed thus:

$$\gamma = f_{\gamma} \cdot f_{O_2}$$

This is the fraction of planets with appropriate surface light (f_{γ}) multiplied by the fraction of planets f_{O_2} whose life-forms use oxygen respiration (a.k.a. breathing or the like).

Let's revisit our updated Drake equation and start plugging in some values to guess the number of star systems in the Milky Way hosting multicellular life:

$$N_{mcl} = N_* \cdot S \cdot P \cdot f_l \cdot \gamma$$

The number of stars in the Milky Way, N_*, is a few hundred billion, so $N_* = 10^{11}$.

Researchers in Hawaii and Ontario, Canada, have estimated that around 1.2 percent of Milky Way stars could host life-bearing worlds. This means $S = 10^{-2}$.

To determine the fraction of planets having suitable conditions for life, P, and the fraction of those with conditions right for the development of oxygen-producing photosynthesis, γ, we can use our solar system as a guide.

Approximately 25 percent of our solar system's thirty-eight

worlds contain abundant liquids. Let's be conservative and re-
duce this number to 10 percent. Since the other terms in P are
likely nearly equal to 1, we obtain $P = 10^{-1}$. One of ten ocean
worlds in our solar system (that is, Earth) has evolved simple life.
So I'm going to take $f_l = 10^{-1}$.

Since only our planet has confirmed photosynthesis, I'll be
conservative and assume that 1 percent of planets receive enough
light for developing photosynthesis. So, $\gamma = 10^{-1}$.

Plugging in values for our terms, we get the following:

$$N_{mcl} = N_* \cdot S \cdot P \cdot f_l \cdot \gamma = 10^{11} \cdot 10^{-2} \cdot 10^{-1} \cdot 10^{-2} = 10^5$$

This means that roughly a hundred thousand star systems can
host multicellular life in a galaxy containing a hundred billion
stars. We are literally one in a million!

But wait! There's more.

We now know that essentially all stars host planets, but we
also know that most planets don't orbit stars! In any given galaxy,
there are many more so-called rogue planets than the star-
orbiting variety on which we live. Rogues fall into two primary
categories: planets ejected from their solar systems and failed
stars. This latter group consists of objects that began collapsing
inside a giant molecular cloud but never entirely formed into
fully fledged stars.

Before we continue, it's time for a change of nomenclature.
Failed stars, a.k.a. brown dwarfs? Rogue planets? I find these
terms, "failed," "brown," "dwarf," and "rogue," to be inaccurate
and overly judgmental. Those little baby stars did not fail at anything. I prefer to call them starlets. And as for rogues, they aren't
aberrant. They were ejected—likely kicked out of their stellar
system by some big bully planet. So, I will refer to starlets and
ejected planets commonly as free planets.

Estimates for the number of free planets in our galaxy range
from tens of billions to trillions. Although the exact number is

highly uncertain, we can be conservative and estimate that there are probably a few hundred billion free planets in any galaxy.

The question here is this: Can free planets harbor life?

Think about it this way: If we moved Jupiter and its moons far from the Sun, would that eliminate the possibility of life on Europa? I don't think so.

Like every phenomenon in the universe, energy is the essential ingredient for life. And for simple, microbial life, geological energy is just as good as sunlight. On a free planet, that energy can come from internal heat sources, not just stellar radiation.

There are two plausible ways a free planet could generate thermal gradients and stay warm enough to support microbial life.

The first is tidal heating. This is frictional heat generated inside a planet or moon as it flexes under the gravitational pull of nearby massive objects. As its shape constantly deforms, heat builds up, fueling internal geothermal energy that can drive volcanism. If this heat reaches a subsurface ocean or reservoir, you've got a potential life-supporting environment. Jupiter's moon Europa likely has tidal volcanoes at the bottom of a global ocean under its miles of ice.

The second is a natural nuclear reactor, and this isn't hypothetical. We've found one on Earth, in Gabon, West Africa. There, a naturally occurring concentration of uranium sustained a self-regulating fission reaction for hundreds of thousands of years, possibly millions. If such a reaction occurred near liquid water, it could create a long-lived warm zone—again, a candidate environment for life.

But what about light?

That's where the story changes. While microbial life can survive and even thrive without sunlight, by feeding off chemical gradients or geothermal vents, multicellular life is a different beast. Complex organisms require a significant amount of energy, and on Earth this energy is derived from oxygen. Free oxygen, in turn, comes from photosynthesis, which requires light.

So no, you don't need light to have life. But if you want forests, jellyfish, coral reefs—or brains—you need oxygenic photosynthesis. And for that, you need photons.

That's why free planets may host microbial life, but probably not civilizations.

How many free planets meet these geological criteria? The truth is, we don't know. We do know that it isn't zero and that there are a lot of free planets—at least as many as there are stars in any given galaxy. Maybe 1 percent of those have the correct geologic conditions for simple life. So, in a galaxy like ours, there should be around one billion, or 10^9, free planets that host simple life.

The equations presented above allow life to develop on any of the universe's water worlds, free planets, or even inside a radioactive comet! For all we know, the weird hydrocarbon lakes and rivers like those on Saturn's moon Titan may be teeming with simple life. As I've said, life may be so hardy and adaptable that water may not be the only liquid upon which life can be based. And since planets tend to be as unique as people, this bodes well for the existence of life throughout the cosmos.

Simple, single-celled life is likely everywhere, but it's locked chiefly under miles of ice, rock, or atmosphere. And even if there are some other unknown energetic pathways to creating widespread multicellular life on these worlds, their inhabitants will never be aware of a larger universe beyond their isolated habitat since, for whatever reason, they cannot stand on their home world and look up and see, quite easily, into outer space. For the same reason, we may never be aware of their existence. This fact answers the riddle of Fermi's paradox, which asks why we have found no clear evidence of intelligent species in our galaxy: Perhaps it's because we're obscured from them and, for now, they're obscured from us.

We would be much more bullish on the existence of intelligent life if we knew that the seeds of life (prokaryotes and viruses) are everywhere around us, locked away under the surfaces of planetary bodies. The molecular precursors to life certainly are; we've found them in meteorites, on comets, and on asteroids. Our problem is that we have not yet developed the ability to directly observe and confirm the existence of non-earthly prokaryotes or viruses, even on bodies in our solar system. But this will not last. Not long ago, we couldn't detect planets orbiting distant stars. Now we do so routinely. And if NASA is able to bring back those Martian samples I mentioned earlier, maybe we'll find the smoking gun we've been seeking.

Perhaps we'll eventually find liquids containing simple life under the ice of Europa, Enceladus, Pluto, or some other solar system crevice. Then perhaps we'd find simple life everywhere, even on worlds orbiting distant stars. Knowing that the "life seeds" that spawn multicellular life and ultimately intelligence are everywhere would mean it's only a matter of time until we find intelligence elsewhere in the universe. We exist. If our existence is possible, then life-forms like ours—multicellular, oxygen-using life—exist elsewhere. Remember what I said about large numbers. They make the possible certain, the rare inevitable.

We're unique in our local corner of the cosmos, but in the vast scheme of reality we're almost certainly not *that* special.

Consider that our observable universe hosts hundreds of billions of galaxies. This means the number of worlds with life may be one billion times one hundred billion, or 10^{20}. That's a universe full of life if I ever heard of one! And this is just in our *observable* universe. As we'll appreciate in later chapters, there's much more to reality than what we can see.

We arrived in the Realm of Life from the Middle Realm, the only of the Nine Realms where life is possible. So, let's take a jaunt back there and address one of those philosophical questions I posed in the introduction. What's happening in the Middle Realm?

What is its *purpose*?

Another preposterous question of the type that scientists don't ask. The universe is not a conscious entity. At least there is no evidence of such. However, fundamental physical processes, such as the motions of light and matter, involve a change to fulfill a purpose. When light changes its location and travels from one material to another, or even from one state to another within the same material, its speed and direction change. We see this all around us every day. A microscope. A telescope. The apparent bending of a spoon in a glass of water. The twinkle of stars.

What determines the path the light takes? *The principle of least time*. Light takes the path that minimizes the time it takes to travel from one location to another. Likewise, the movement of matter and light is governed by the more general *principle of least action*. This obscure physical quantity called the action, which has units of time-energy, is minimized as matter moves from one point to another.

If I had to come up with a name for the principle that governs how matter changes in the Middle Realm, I'd call it the *principle of least energy*. As I noted at the end of the previous chapter, the Middle Realm evolves to remove as much energy from matter as possible. Where does the energy go? It becomes light. Look up at the night sky, and you'll see that points of light cover it everywhere. If you're in a sufficiently dark location, a swath of diffuse light stretches from horizon to horizon. These are the hundreds of billions of stars of the Milky Way. All represent incomprehensible numbers of protons whose energy has been (and is being) converted to light. Most of that light will travel through space forever, never to be incorporated into matter again. And some of it will be reabsorbed by matter, but not by protons. Once again, electrons will eventually absorb this excess energy and reconvert it to light. This process has been unfolding for billions of years. And it will continue for trillions of years to come. Over time, the amount of energy that exists in the universe, in association with matter, steadily decreases, and the amount of light in the universe steadily increases.

Yoda was right: "Luminous beings are we, not this crude matter."

And amid this wondrous Middle Realm are we and all of our living brethren. Bathed in proton-to-photon-to-electron-to-photon energy; energy we store like batteries in the configurations of our electrons.

Life *is* energy, and energy *is* life.

COSMOLOGICAL REALM

The fact that we live at the bottom of a deep gravity well, on the surface of a gas-covered planet going around a nuclear fireball 90 million miles away, and think this to be normal, is obviously some indication of how skewed our perspective tends to be.

—Douglas Adams

If the Middle Realm is the realm of the conceivable, and the Realm of Life stretches from complex organisms like us to life-forms so basic they blur the line between chemistry and biology, then the Cosmological Realm is the realm of *space-time*. It is the realm of discarding your notions of distance. And for that matter, you can also throw away your notion of time. They don't work here. That phrase you've heard repeatedly, "The age of the universe"? There is no such thing. Energy can be neither created nor destroyed? That's out the window, too. An object at rest? Nope. Can't happen.

The Cosmological Realm represents the vast, large-scale volume that we refer to as our universe. Its range extends from the size of a typical galactic arm, around twenty thousand light-years, to the size of the entire observable universe and beyond. There is little in our everyday lives that prepares us to think on cosmological scales. Even when we go ambitiously large in our science fiction (allowing me to mention Marvel's Galactus the World Devourer again), we still don't think nearly big enough. But as big and weird as it may be, our minds *can* grapple with this realm and develop ways to see it.

I still remember the feeling that came over me the first time I saw the night sky in 3D after an evening of teaching observational astronomy at Stanford's Student Observatory. That night, the Moon, planets, and setting Sun transported me out of my earthly existence and into space. In preparing for my classes, I had recently learned a few numbers: the size of Earth, the sizes of all the other planets relative to Earth, and their distances from Earth. As I stepped out of the observatory and looked up, I could see the Moon as a planetary body one-fourth Earth's size and around sixty Earth radii away. I could intuit the Moon's actual size, and it was gigantic! I then found myself looking beyond the Moon to Jupiter on the opposite side of the Sun, eleven times Earth's diameter, orbiting the Sun five times farther away than Earth. Jupiter now appeared monstrous as I internalized this planetary scale. But still, I remained in the Middle Realm. I would not venture to the Cosmological Realm until that fateful night on the Northern California mountaintop when I saw Andromeda with my naked eye.

Seeing Andromeda broke something open in me. Until then, I had been stretching my mind across the solar system, scaling up from the Moon to Jupiter and beyond. But that moment shattered the illusion that space was just a bigger version of what I already understood. The scales weren't just vast, they were fundamentally different. Distance and time started to feel wrong. My familiar notions of movement, size, and sequence no longer applied. The night I viewed Andromeda, I stepped out of the Middle Realm and caught my first glimpse of something stranger: the Cosmological Realm.

At this point, I would love for you to see the Cosmological

Realm in 3D. To do this, let's start by imagining ordinary matter and work our way up. This means returning to the simple atom.

If you look at atoms one way, you find they're not dense. The nucleus—an atom's protons and neutrons—accounts for 99.99 percent of an atom's mass, but only one ten-trillionth of a percent of its volume! That's 10^{-15} or 0.000000000000001. The volume is determined by the boundary of the atom's electrons, which zip around the massive but small nucleus at a great distance. How great? Well, if the nucleus were a car and the electrons were loaves of bread, then the bread would zip around the vehicle at a distance of two hundred miles!

We are taught that ordinary matter is mostly empty space, but this is really empty-*ish* "electron space." As you'll see in the Quantum Realm, electrons are not the tiny spheres we typically imagine. They're some weird quantum wave thing that arranges its physical form based on where it is and what it's interacting with. In a hydrogen atom, the volume that the electron defines is *one quadrillion*, or 10^{15}, times larger than the nucleus! And because of the weirdness of the quantum realm, this electron inhabits this entire space at all times, at least until our hydrogen atom is forced to interact with something. The "empty space" idea isn't a bad analogy, given that the volume of an atom is so much larger than the mass concentration of the nucleus, but it isn't empty per se. It's the domain of the electron.

Of course, this is not just true for hydrogen atoms; it's true for all ordinary matter. Consider a one-centimeter cube of pure iron—one of Earth's most abundant elements. Iron nuclei are regularly spaced in a crystal lattice, and the distance between each nucleus is roughly the diameter of the atom itself. Picture the iron. The tiny nuclei of these iron atoms, which hold practically all the iron's mass, are separated by a distance that is a *couple of hundred thousand times* their size from each other. (If humans

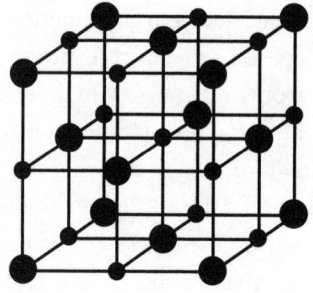

Fig. 9: CRYSTAL LATTICE
Atomic arrangement in a crystalline solid. These patterns determine material properties and were key in the early formation of rocky planets.

were distributed like iron nuclei, they'd be more than a hundred miles apart!) The nuclei are like little pinpricks of mass held in a swarm of electrons. Iron is a hard and heavy substance, but at the atomic level its "electron space" is occupied by a field of near-massless electrons!*

Let's move up a few clicks. In the Cosmological Realm, the basic units of mass aren't atoms or molecules; they're stars and galaxies. Let's start with stars.

On average, stars are light-years apart but only a few light-seconds in size (meaning it takes light only a few seconds to travel a distance equal to a star's diameter). Since there are around thirty-one million seconds in a year, a few light-years' separations translate to stars being *tens of millions of times* their size apart from each other. This spacing is a hundred times *greater* than that for our iron nuclei, which were a couple of hun-

---

* But if it is empty in terms of space, it is full in terms of what it takes to make our small cube of iron: A one-centimeter cube of iron contains 10^{24} iron atoms. That's a 1 followed by 24 zeros. Imagine how many iron atoms it must take to make up Earth's core!

dred thousand times their size apart. The realm of stars is emptier than that of atoms by several orders of magnitude.

Which brings us to galaxies. Continuing our reasoning, how far apart in terms of their size do you expect galaxies to be?

You're probably thinking, "Really freaking far, Dr. O.!"

Well, I've set you up again. The truth is that on cosmological scales, the universe is relatively *full*.

As you know by now, the nearest large galaxy to the Milky Way is Andromeda, which is only around thirteen times the Milky Way's diameter away. Heck, the entire observable universe is only a million Milky Ways across! If galaxies were distributed like stars, Andromeda would lie far beyond the edge of the observable universe. It turns out that galaxies are really, really close together. If we scaled the Earth-Moon system to the dimensions of the Milky Way–Andromeda system, the Moon would be more than twice as close to Earth as it is now.

If you want a mental model of our galaxy-packed universe, you're in luck. There exists a collection of objects you're very familiar with whose constituent structures are separated by roughly ten to a hundred times their size, just like galaxies. Air molecules!

Guess what volume constitutes roughly four hundred billion air molecules (the number of galaxies in the observable universe)? One cubic millimeter of air.

Think of that! To a being one thousand times the size of our observable universe, our universe of galaxies looks just like how a cubic millimeter of air looks to us.

Astronomers working in the Cosmological Realm routinely look at our universe as a whole. To simplify matters, we use a rule known as the cosmological principle, whose main consequence is that our universe is both homogeneous and isotropic. "Homogeneous" means that it comprises the same stuff everywhere, and "isotropic" means there are no special directions. Put differently, matter in the universe is uniform, and the universe has no center and no edge.

Why galaxies are organized the way they are in our universe is a matter of time and statistical probability. (And yes, I've been writing "our" universe. There may be other universes! We'll get to those.) Their distribution could have been much denser (as they once were), like water or rock molecules. If that were the case, our universe would have been a much more violent and chaotic place to eke out a living. Conversely, the distribution could have been not dense at all, like the sparsest corners of the cosmos. Here, our universe would have been a cold and lonely place.

But as luck has it, we live here and now in our universe.

And yet, on the largest scales, our universe—the one that made our existence possible—is governed by rules that are anything but familiar. The Cosmological Realm is weird, not just because of its vastness, but because the dominant processes that drive its evolution follow rules that are opaque to our everyday experience. Energy flows differently here. Gravity behaves differently. And the reason for all of that—the foundation beneath it all—is the outsized role that *space-time* plays in shaping this realm's behavior. To understand the Cosmological Realm, we don't just need to scale up our imagination; we must reshape it entirely.

Perhaps you have some hazy concept of space-time, or maybe it sounds like a made-up word jumble whose definition seems impenetrable. Maybe you've heard it described as the interaction of space and time, or assumed it's some kind of cosmic coordinate grid. Some imagine it as a stretchy fabric on which everything sits, or a background stage where the universe plays out. But none of those metaphors quite capture what space-time is— and why it rewrites everything we thought we understood about motion, gravity, and even reality itself.

We physicists are not always great with words. "Space-time"

has a nice ring to it, but it can sometimes be confusing to discuss space and time, and space-time. As we will soon appreciate, space is related to time, and time is related to space. But space-time is *not* this relationship. It is its own thing. There is space, there is time, and there is space-time.

Picture space a vast, empty 3D grid that stretches in all directions. You can place objects in it—a star here, a planet there—but nothing moves. Without time, there is no change.

Now picture time: a straight arrow running forward. Along that arrow, you can place events: birth, explosion, orbit, collapse—each moment ticked off like notches on a ruler. But without space, there's nowhere for anything to happen.

Now combine them. Place grids of space across that arrow of time. But instead of space and time staying flat and straight, imagine they warp, curve, expand, compress, and wave. A massive star pulls the grid downward, creating a dent. A black hole stretches it into a funnel. A beam of light moves not in a straight line but along this distorted surface. This is space-time—a single entity where motion, gravity, and even the flow of time are all determined by the shape of a four-dimensional space-time framework.

Space-time does not possess speed, yet it moves everywhere. You can't touch space-time, but it touches everything. You can't see space-time, but the world is visible thanks partly to space-time. Space-time exists throughout our universe, everywhere, at all times. And it's not just a background. It does stuff. It curves, stretches, waves, expands, contracts, and, under certain conditions, is irresistible. It's not a reach to say that space-time *is* our universe. Everything that happens in our universe occurs somewhere and somewhen. The combination of all possible wheres and whens is space-time. If you can understand space-time even a little, you can begin to understand the mysteries of the Cosmological Realm.

Let's give it a shot.

First, as its name suggests, where it concerns space-time, space and time are two dimensions of one phenomenon—one space-time. This is true at the largest scales and the most minuscule.

Second, there is a minimum and a maximum speed at which anything can move through space or time. The minimum speed is zero. The maximum speed we denote with the letter c, which is better known as the speed of light. But this is a misnomer.

This constant speed isn't just a property of light; it's fundamental to the structure of space-time itself. It sets the ultimate speed limit for everything in the universe; no information, matter, or energy can travel faster than c. It's the speed at which light travels through a vacuum, yes, but it also describes how gravity works, how particles interact, and how the universe expands.

Third, similar to how we independently move vertically and horizontally, we move separately through space and time, though *not* independently.

Fourth, everything (including you, your dog, and your house) moves through space-time at the top speed, c, at all times, not through space alone or through time alone, but through space-time. That we are all moving at the speed of light is counterintuitive, but it's true. When it comes to space-time, the speed is split between the two dimensions of space and time: The faster one moves through space, the slower one must move through time, and vice versa.

At this point, I've used the term "space-time" over and over. My guess is that to have an image or idea to grasp on to, you'll probably spend much of this chapter trying to connect my description to visuals you've retained from TV science programs or online media. Unfortunately, the truth is that we physicists spend years and decades of our lives attempting to visualize space-time. It's hard.

Now, I know I promised not to drag you through too much math. But bear with me for just a moment, because this next part is surprisingly simple and powerful. The main idea regarding space-time is based on geometry's most famous theorem: the Pythagorean theorem. That's the one for right triangles that anyone who has ever attended a middle school math class recites as "*a* squared plus *b* squared equals *c* squared." Here's what that looks like:

$$a^2 + b^2 = c^2$$

The Pythagorean theorem (dating back more than twenty-five hundred years!) states that in a right-angled triangle, the square of the length of the hypotenuse, *c*, is equal to the sum of the squares of the other two sides that form the right angle, denoted *a* and *b*. (The *c* here is not to be confused with our *c*, the speed of light.)

This is important because general relativity (GR), Einstein's great theory describing space-time, is fundamentally a theory of geometry. And the simple Pythagorean equation of triangles is the gateway to understanding how distances work in GR's curved space and, ultimately, how gravity and space-time are one and the same.

First, let's remove the specificity of triangles and move to a general two-dimensional plane with coordinates *x* and *y*. We also define the length of some randomly oriented line as *s*.

We can then write:

$$s^2 = x^2 + y^2$$

Now let's add a third dimension to describe a volume. We now need three coordinates, so we will call this new dimension *z*, as is customary. The length of the line *s* in this 3D space can now be written as:

$$s^2 = x^2 + y^2 + z^2$$

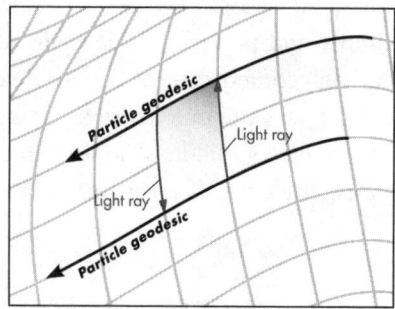

FIG. 10: SPACE-TIME SURFING
*An object (or astronaut) moving through curved space-time, illustrating the equiva-
lence principle and free fall as weightlessness. Free-falling objects follow geodesics—the
"straightest" possible paths in curved space-time.*

Then Einstein comes along and informs us that we don't live
in a three-dimensional universe after all. We live in a four-
dimensional universe; the fourth dimension is time, which we can
denote as t. But we can't just add t^2 to the equation as we did with
the z dimension. The reason is that we can only add those things
that are the same. For example, two apples plus three apples equals
five apples. But two apples plus three polar bears are just two apples and three polar bears. We need to turn time into distance in
Einstein's new four-dimensional space-time. This is accomplished
by multiplying time by the speed of light, c, to get ct as our new
coordinate. We can now calculate a space-time interval as:

$$s^2 = (x^2 + y^2 + z^2) - (ct)^2$$

Or, if you prefer a slightly less technical version of this equation:

$$(space\text{-}time\ interval)^2 = (space)^2 - (speed\ of\ light \times time)^2$$

But let's stay with the more technical version for a minute. We
can clearly see how this four-dimensional "space-time interval,"

s, deviates from standard 3D space. The time coordinate (*ct*) has a negative sign in front of it, opposite the spatial coordinates, which means they push and pull in different ways when we calculate the interval. This has profound consequences for how events in the universe are interconnected and how we understand concepts such as causality, the speed of light, and the very nature of reality.

In the Middle Realm, we think of events separated in space (distance) or time (duration). But in Einstein's 4D space-time, there are three types of separations: time-like, space-like, and now light-like.

Lightning strikes offer a vivid way to explore space-time separation types. Let's break it down through three different lightning scenarios, each one revealing a distinct kind of separation between events.

First, imagine lightning strikes the same tree twice, a few seconds apart. These two events are time-like separated. The first bolt alters the environment—ionizing the air, changing the electric field—and those changes can directly influence the second strike. That's because time separates these events more than space does. A beam of light could easily travel the short distance between the two strikes long before the second one happens. This means a causal connection is possible. Every observer, no matter how fast they're moving or in which direction, will agree on the order: first strike, then second. The events lie within each other's light cones, and causality is preserved.

Now flip the script. Picture two bolts striking at the same instant, but in cities a thousand miles apart. It would take light almost 5.5 milliseconds to travel that distance. The bolts last for only 0.2 milliseconds. By the time light from a lightning strike travels a thousand miles, the original discharge channel has long since cooled and dissipated. These are space-like separated. No signal, not even light, can travel from one location to the other in the available time. There's no way for one event to influence the other. And here's the kicker: Different observers, depending on how they're moving, won't even agree on which strike happened first. Some might say City A's strike came first, others might say City B, and some might call it a tie. These events live in each other's "elsewhere" region—a part of space-time that is causally disconnected. They exist in the same universe but are isolated slices of reality.

Then there's the third case: Lightning strikes a mountaintop, and an observer down in a valley sees the flash exactly when light from the event reaches their eyes. This is light-like separation. These two events—the strike and the observation—are connected by a beam of light, traveling along the precise edge where the contributions to space and time separation perfectly balance. Physicists refer to this path as a null geodesic. It's the route light takes through space-time, sitting right on the boundary between what can influence what, and what can't.

These three types of separation—time-like, space-like, and light-like—expose the deep geometry of space-time. The lightning strikes themselves are just points where something happens. But the relationships between them define how causality operates, how influence moves, and why the speed of light sets the ultimate limit. Every pair of events in the universe falls into one of these categories. That simple fact shapes the architecture of reality itself.

The fact that time and space are treated with opposite signs in our space-time interval equation ensures that this cosmic speed limit—the speed of light—is never broken. This is crucial for maintaining the rule of causality: Causes must *always* happen before effects. In our everyday experience, this seems obvious: You flip a switch, and then the light turns on; you throw a ball, and then it hits the wall. But in the vastness of space-time, it's the geometry of space-time itself that guarantees causality holds.

This brings us to the next important point: The relationship between time and space differs for everyone. Two people moving at different speeds might disagree about whether two events occurred simultaneously or how far apart they were in space. This is known as the relativity of simultaneity, illustrating the profound interconnection between space and time. Einstein provided a now-famous example: Imagine sitting in the center of a fast-moving train car, holding a lantern. A door with a light sensor is at each end of the car. When light of a particular color reaches the door sensor, the door opens. It just so happens that your lantern produces the correct color. You turn on the light. Since you're sitting in the center of the train car, the doors are equally distant from you. When you turn on your lantern, you will see the light travel in all directions, eventually reaching the two doors simultaneously and causing them to open simultaneously.

However, someone standing outside the train on the ground would see something different. They see the rear door of the train car approaching the light and the front door moving away from the light. Consequently, they see the rear door open first followed by the front door.

Neither perspective is wrong; it's just that in space-time the way space and time mix depends on how fast you're moving. This idea only makes sense because of the space-time equation's opposite signs for space and time.

Even though people moving at different speeds might disagree about the timing (t) or location (x, y, z) of events, they all agree on something more fundamental: the space-time interval, s, between events. This interval is an invariant quantity, meaning it remains the same regardless of who measures it. The different signs for space and time ensure that the overall interval remains constant even though space and time coordinates change depending on your motion. This profound truth about the universe allows the laws of physics to hold true for everyone everywhere all at once; it doesn't matter how fast (or slow) anything is moving, nor does it matter which direction anything is traveling.

The fact that space and time have opposite signs in the space-time interval equation transforms how we think about the universe. It defines whether events can be causally related, sets the ultimate speed limit for the transfer of information, and explains why different observers experience time and space differently while still agreeing on the essential physics of what happens. This interplay of space and time provides us with a coherent and consistent understanding of the universe—one where the speed of light serves as the ultimate gatekeeper of reality.

Here is one example. If we divide both sides of the space-time interval equation by time squared (t^2) and rearrange terms, we obtain a new form of the Pythagorean theorem but in terms of our speeds through space, v_s, and through time, v_t:

$$v_s^2 + v_t^2 = c^2$$

This equation tells us that at all times, everything in the universe moves at the speed of light, c, through space-time. But, since all measurements of space and time are relative, different observers will measure different values for our speed through space and time, respectively, depending on their motion and local gravity. Since the speed of light, c, is fixed (meaning c^2 is also fixed), the sum of v_s^2 plus v_t^2 must always equal the same number: c^2. Therefore, if v_s decreases, then v_t must increase, and vice versa.

As Einstein understood, time is relative. The more significant the difference in space-time curvature between two locations, the greater the difference in v_t—the speed of time—between them. And the more significant the difference in v_s for two objects, the greater the difference in v_t between them. Each of these examples says the same thing: *the greater the energy difference, the greater the difference in the speed of time.*

What about light, not just visible light, but all forms of light? How does light experience time? A fundamental property of light is that its speed in a vacuum, c, is the same for all observers, regardless of their motion. If an observer is traveling at half the speed of light, it still observes light as traveling at the speed of light, c; if an observer is traveling at 99 percent the speed of light, it still observes light as traveling at the speed of light, c. Given this, you might think that light exists perpetually in a state where its velocity through space is $v_s = c$, and its velocity through time is $v_t = 0$. This reasoning, combined with the principle of relativity, which requires spatial distances to contract toward zero as an object's speed approaches c, has led to a widely held misconception: that light does not experience space or time.

Here's how that argument is often framed: Imagine a photon traveling to Earth from a galaxy billions of light-years away. According to this view, the photon would describe its journey as instantaneous. It is emitted, and then, without any passage of time or travel through space, it is absorbed here on Earth. From the photon's perspective, no time elapses, and no distance is crossed. While humans perceive the journey as spanning billions of years of time and many light-years of distance, these dimensions supposedly collapse to zero for the photon. The story begins and ends instantly: "You are, and you have arrived."

But there's a critical flaw in this line of thinking. While it is true that time durations and spatial distances are relative, meaning they change depending on the observer's frame of reference, an essential concept is tied to an object's rest frame. When we ask, "What is the experience from the photon's perspective?" we imply the existence of a special frame: the one in which time flows at its maximum rate for the entity in question—its rest frame.

Here lies the paradox: *Light has no rest frame.* Unlike objects with mass, which can exist at rest relative to some observer, light always moves at c. No matter the observer's motion or gravitational conditions, they will always measure light traveling at the same speed. This universal property makes it impossible to define a rest frame for light. Without a rest frame, the concept of time or distance as experienced by light becomes meaningless. In essence, the question of how light experiences time is unanswerable.

■ ■ ■

Space-time is challenging to comprehend because we don't notice it in everyday life, even though it undergirds our very existence. But we do see one manifestation of space-time nearly all the time. We notice it through the common phenomenon known as gravity.

Here we are back to Newton—back to falling. But what exactly is falling? What is gravity?

Well, primarily, gravity is a manifestation of space-time. More specifically, gravity is a manifestation of space-time's *curvature*.

We can envision the geometry of space-time with a mental experiment. Consider what would happen if two objects—let's say cute little polar bear cubs—were a certain distance apart and moving in the same direction through outer space. Suppose we ignore their mutual gravitational attraction (two cubs have very little mass and exert practically zero gravitational force on each other). In that case, our baby bears will continue forever in their straight lines, maintaining their distance and never crossing paths.

But their separation would decrease if our cubs approached a planet or star. The closer they travel to this planet or star, the more their straight-line paths will curve. If they could continue on their paths unabated, their paths would eventually intersect near the center of the planet or star.

Of course, they'd die in a fiery entry into the planet's atmosphere or star in real life. And if the planet doesn't have an atmosphere, well, splat!

Albert Einstein realized that all inertial paths in the presence

of a spherical planet or star are curved in such a way as to intersect near the center of the body.

Space-time curvature exists around planets, stars, and galaxies, and this space-time curvature creates gravity. Without space-
time curvature, particles in the universe would fly around in
straight lines, caroming off each other like billiard balls in a random dance of chaos. Nothing would settle into place. There
would be no hope of life ever starting, because there would be no
stars or planets.

▪ ▪ ▪

Space-time exists everywhere throughout our universe at all
times. It exists *locally*, around Earth, in our solar system, and
globally on cosmological scales. In fact, on the largest scales,
space-time curvature determines whether we have a universe at
all.

To understand how this works, we must distinguish between
the local space-time curvature around a planet, star, or galaxy (or
you!), and the global curvature of our universe.

We have to regard space-time from the perspective of the
Cosmological Realm.

Consider Earth an analogy for our universe. Looking at
Earth up close, you see significant variation on its surface. You
see mountains, valleys, gullies, hills, trenches, and sinkholes; you
see expanses of emptiness and "flat" prairies and oceans; you see
greens, blues, ochers, and reds. Locally, it's got a lot of character.

The universe is no different. Local regions in the Cosmological Realm possess a similar local space-time curvature topogra-
phy. There are wells and divots around stars and planets; there
are great intersecting sinkholes around galaxies. At these scales,
space-time is a never-ending succession of rolling hills and by-
ways, all working together to ascribe motion to the universe's
constituents.

Pull back and you see something different. At a distance, Earth appears as a smooth blue-and-white ball that bulges at the equator and is slightly flattened at the poles. As I pointed out at the beginning of this chapter, when you step back and look at the universe, it, too, appears to be uniform. It is the same everywhere—it's homogeneous—and has no center or edge—it's isotropic. The same holds for space-time. When you pull back far enough, it's not bumpy; it has zero curvature. It's flat. Since, in some very simplistic sense, the shape of space-time is the shape of our universe, our universe is flat, too.

So basically, the "Flat Earth" people aren't thinking big enough. Flat Earth does not exist. But Flat Universe does!

Let's break this down. The global curvature of space-time is determined by the total density of everything that exists. Not the total *amount* of everything—the total *density*.

Density is how much stuff there is in some section of our universe. To double the density of a region, either you can double the amount of material and keep the volume fixed, or you can squeeze the same amount of material into half the volume. Both regions would have the same density.

But there's something weird going on in our universe. *Space-time is expanding.*

That's correct: The volume of the universe is ever increasing. We would expect the density of stuff in our universe to change with it. But this doesn't happen. It turns out that the amount of stuff is not fixed; it's *also* increasing. (We can thank energy for that!) If the amount of stuff in our universe evolves just the right way, it balances out the expansion of space-time. This keeps our universe's density more or less constant, which in turn keeps global space-time curvature flat.

How can I say this with any authority? Amazingly, humans have used our imaginations to devise techniques that measure our universe's curvature. The concept is straightforward. Light follows the shape of space-time; it has no choice. Like matter, light is trapped in space-time. If space-time curves, then the light in it changes direction, too. And if space-time stretches—as it does throughout the universe—the light stretches, too.

Since light must follow the shape of space-time, if a source of light of known size or shape is visible from a great distance, those characteristics would be systematically modified if the light traveled through globally curved space-time. The source would look smaller if space-time curved one way, larger if space-time curved the opposite way, and unchanged if space-time didn't curve at all. Well, we've done the observations, and we have found that our universe has been measured to have zero curvature with a precision of better than half a percent.

Having zero curvature at the current age of our universe means that the density of matter and energy at the beginning of our universe had to have been highly precise; it had to be fine-tuned way back then to have zero curvature now, 13.8 billion years later. To have it end up this way after billions of years of expansion is like trying to keep a sharpened pencil perfectly balanced on its tip for 13.8 billion years—extremely difficult! Any tiny imbalance at the start would have grown more prominent over time. A universe that started just a bit denser would have collapsed back in on itself, while one that started a bit less dense would have expanded so fast that galaxies wouldn't have formed.

On the scales of the Cosmological Realm, it helps to look at space and space-time independently. Time does not expand, but the expansion of space can be thought of like the flow of a river. And galaxies can be thought of like tiny tadpoles in a moving current. Imagine standing on the bank of the river and observing these tadpoles swim in waters of varying speeds. Near the riverbank, the water interacts with the land and flows much more slowly than at the river's center. The tadpoles near the bank, where the water is almost still, can flap around with their tails and move at will in their search for food. We can easily see the tadpoles' speed relative to the water. But for tadpoles closer to the river's center, the water's flow dominates their motion. They're still flapping around doing their thing. But we see them swept downstream. This is similar to how we observe the motion of galaxies in the expansion of space.

Even though galaxies have their independent motion, which we call their peculiar motion, the motion of a galaxy far, far away is dominated by the speed of expanding space. Astronomers refer to this flow as the *Hubble flow*. And we call the effect this flow has on galaxies' peculiar motions *expansion drag*.

Eventually, galaxies lose their kinetic energy and reach a standstill relative to space's expansion. Believe it or not, for the most part, galaxies don't move; they're stationary relative to space's expansion current. Like the tadpole in the middle of the river, they're just along for the ride.

A key concept to understand about a ride powered by expansion is that the farther away something is, the faster it moves away. They are directly proportional. Objects twice as far away move away twice as fast. Objects three times as far away move away three times as fast. And so on.

Imagine grabbing a rubber band at both ends and holding it in front of your body at full, unstretched length. Now, in one second, stretch it to twice its original length. It's easy to see that the two ends of the rubber band moved away from each other at a speed of one full length per second. But how fast did each end move from the center of the rubber band? Only half that speed!

In more detail, the ends of the rubber band went from being a single length apart to being two lengths apart. At the same time, the center moved away from either end by half that distance. And the regions of the rubber band that are very close to the ends? They hardly moved at all.

The exact same thing holds for our universe. Galaxies near us, like Andromeda, which I witnessed that night in California, are not carried away from us by our universe's expansion. Andromeda and the Milky Way are so close that gravity is pulling us together. In fact, in about four billion years, Andromeda and the Milky Way will collide, and about six billion years after that their merger will be complete. Their corresponding supermassive black holes will eventually spiral around each other and also collide, sending strong gravitational waves across our universe. A new galaxy will be left in their wake—Milkomeda according to the nerds, but the cool cats know it's the Hakeem Galaxy—and physicists believe that it will not be a spiral galaxy but an elliptical one. Earth will be long gone at that time. But if we're still around living on (or in) some other body, it will be quite the show!

But most galaxies in the universe are moving away from us, just as we are moving away from them. (Besides Andromeda, there are a few other galaxies in our Local Group that are also close enough to us that they are moving toward the Milky Way and not away.) And believe it or not, at great distances, galaxies are moving away from us *faster* than the speed of light. This is all due to the expansion rate of space-time.

You may be thinking, yet again, "Whachoo talkin' about, Dr. O-lis? I thought you said nothing could move faster than light." This is true, nothing can move faster than light *through space-time*. But some things do *effectively* move faster than light, and it's not difficult to visualize how.

Imagine you are standing in the middle of an infinitely long airport terminal. For this analogy, you are the Milky Way. Just next to you is a walkway that also extends infinitely into the distance. This walkway is not like the moving airport walkways we're familiar with. It doesn't move like a treadmill, but instead stretches like the rubber band I mentioned earlier. And not just that, its stretching action accelerates over time. This walkway represents the expansion of space-time.

Somewhere on this walkway is another person-galaxy. This person is stationary relative to her spot on the walkway. But remember, the walkway she is standing on is stretching away from you, and the farther away she gets, the faster she moves relative to you. Since this is an infinite walkway, points that are very, very distant can (and do!) move faster than the speed of light *relative to you*.

Even though this person-galaxy is stationary, she eventually reaches a point where she is moving away from you at velocities greater than the speed of light. Regardless of how fast or slow the stretching occurs, this will always be true. Because space-time is expanding in a volume that is effectively infinite (and may actually be infinite), there will always be a point at some distance from you that is "moving" faster than the speed of light.

Of course, our position in the cosmos is not unique: The universe is isotropic (it is the same in all directions, and no direction means anything special). Which means that you're not really standing still in the middle of the terminal, but rather when you look down, you realize that you too are on a stretching walkway, and you too are moving away from the other person-galaxy. Relative to your walkway, you are also stationary. Relative to the other person-galaxy, you are now moving away from her at velocities that greatly exceed the speed of light, even though nothing in this system—not you, not the other person-galaxy, and neither of the stretching walkways—is itself moving faster than the speed of light.

Unlike "normal" acceleration on Earth, which describes how speed changes over time (usually denoted as meters/second squared), expansion at cosmological scales is a kind of acceleration that describes how speed changes over *distance* (usually denoted as kilometers/second/megaparsec, or km/s/Mpc, where 1 megaparsec is about 3.25 million light-years). Instead of increasing their speed over time, objects in the Cosmological Realm have relative velocities that increase with distance. In the current state of our universe, the expansion rate is roughly 70 km/s/Mpc. Think about that; this means that every 3.25 million light-years or so the expansion rate increases by about 157,000 miles/hour.

There is a characteristic distance at which the expansion of space-time carries objects away from us at the speed of light, and we physicists call it the Hubble sphere, after the great American physicist Edwin Hubble, who first showed the linear relationship between galaxy distance and redshift, confirming that the universe is expanding. Objects inside the Hubble sphere are moving away from us at relative velocities below the speed of light, and objects outside the Hubble sphere are moving away from us at relative velocities greater than the speed of light; all of this is due to the continuing and accelerating expansion of space-time. Currently, the edge of the Hubble sphere is located about fourteen billion light-years away from us here on Earth. (To some other being on some other planet on this other far end of the universe, we are at the edge of their Hubble sphere. It is all relative.)

Beyond the Hubble sphere is another characteristic distance called the cosmic event horizon. This represents the maximum distance from which light emitted now will eventually reach us in the distant future. Light (or any other signal) emitted from objects that lie beyond this boundary will never reach us, no matter how much time passes. At the moment, the cosmic event horizon lies a bit beyond the Hubble sphere at sixteen billion light-years away. It is like a bubble enclosing a bubble.

But there's more! Far beyond the cosmic event horizon lies a third boundary, known as the particle horizon. This is the maximum distance from which light has had time to reach us since the beginning of the universe, 13.8 billion years ago. This horizon represents the edge of the observable universe, essentially defining how far we can see in space due to the finite speed of light, the universe's age, and the expansion of space-time. And believe it or not, even though the universe is "only" 13.8 billion years old, because of the accelerating expansion of space-time, we can see objects that are now much farther away than 13.8 billion light-years!

This is where things get really interesting. Even though some galaxies are beyond today's cosmic event horizon, we can still "see" them because they emitted light billions of years ago, back when they were *inside* our cosmic event horizon. The dynamics of space-time expansion create this effect, allowing us to see light from galaxies that are no longer within the boundaries of our cosmic event horizon.

FIG. 11: COSMIC HORIZONS
Depiction of our cosmic visibility limits: the Hubble sphere, cosmic event horizon, and the particle horizon.

If you're struggling with this, imagine that our person-galaxy in our infinite airport terminal flicks on a flashlight pointed at you, the Milky Way. When she turned on her flashlight, she was within your cosmic event horizon, and neither of you at that moment was riding space-time expansion that exceeded the speed of light. Over billions of years, the walkways continued to accelerate, carrying the person-galaxy sufficiently far away that she crossed our cosmic event horizon (at this point, you also cross her cosmic event horizon). If she flicks on her flashlight now, the light will never reach you since she is beyond a point where light could ever catch you. But the light from her original flashlight flick is still out there, and it is still on its way toward you. Since the space it's traveling through is already inside your cosmic event horizon, it eventually reaches you even though the source of that light is now long gone and will never, ever be seen again by you.

Our particle horizon is the ultimate edge of what we can see in the universe. It is the maximum distance from which light has had time to reach us since the beginning of time. As the universe ages, this particle horizon expands, just like the Hubble sphere and the cosmic event horizon, but it does so in a slightly different way.

As time passes and light from more distant parts of the universe continues to travel toward us, the particle horizon grows larger. It's like watching a movie that gradually reveals more and more of the scene: With each passing moment, you see farther and farther into the cosmic landscape and closer and closer to the beginning of the universe. Today, this horizon is about *46 billion light-years away*, much larger than the universe's age of 13.8 billion years would suggest.

How's that for trippy?

By the way, space-time doesn't just curve and expand; it also twists, oscillates, and sucks.

Go outside on a cloudless night, and outer space will start

sucking the heat right out of you. Hold up an umbrella as a person would if they were trying to block the Sun's heat, and you will reap the opposite result. Cutting yourself off from outer space prevents it from sucking the heat out of you, so you stay warmer.

The oscillations in space-time are caused by colliding and merging cores of dead stars and black holes. We call these excitations gravitational waves. These oscillations pass through stars, planets, Earth, and you, all unimpeded. They cause all distances to be slightly modified as they pass by. So you and everything around you get bigger and smaller when these waves pass through you. This may sound wild, but these waves exist; we've measured their oscillations with gravitational wave observatories called LIGO and Virgo. We've heard the chirps of merging black holes up to a hundred times the mass of the Sun. We've observed pulsars to measure the loud and constant background rumble of supermassive black holes millions of times the mass of the Sun colliding from the dawn of galaxy birth.

These more esoteric manifestations of space-time's dynamics have been invisible to humans primarily because we are so small. Our minuscule size biases us and often closes our eyes to the universe's actual reality. For example, we think the speed of light is fast. But as I noted earlier, the speed of light is fast to *us*. To galaxies, the speed of light does not seem very fast at all. Consider that it takes light more than two million years to reach us from the *nearest* galaxy. In the Cosmological Realm, most changes occur before they can even be observed. There is still cause and effect, but often we start with the effect and then have to suss out the cause.

Combining light's relatively slow speed at cosmological scales with the dynamic nature of space-time yields our everyday experience of distance as null and void in the Cosmological Realm. If you wish to measure the distance to some object, a signal must traverse the distance between the two locations. As we now understand, in cosmologically expanding space, very distant motions occur faster than one's ability to measure them. Consequently, whenever one measures the distance to a cosmological object, it's no longer there. And the farther away a thing is, the weirder and more noticeable the effects become. For example, very distant galaxies can appear to be closer to us than they truly are. It's like that warning on automobile side-view mirrors: "Objects may be closer than they appear." But in the case of the Cosmological Realm, it's the opposite; the most distant galaxies may be *farther* than they appear.

Imagine standing outside a house with your nose against its wall. The house will block your entire field of view. Now imagine yourself walking backward away from the house but still facing it. At a certain point, you can see beyond the home's wall when you look to your left and right. As you continue walking backward, the house will block less and less of your view. Walk back far enough, and the house will eventually appear as a small dot on your horizon.

What I have just described is "normal" to us and intuitive. It does not take a giant leap of imagination to realize that you could, in principle, determine the distance from you to the house based on how much of your view it blocks (how big it appears).

In the Cosmological Realm, we can measure distances similarly. However, there are two significant differences to consider. First, in our example of measuring distances to the house by using its apparent size, we didn't need to account for the time it takes light to travel from the home to your eyes. On Earth, this time delay is negligible because the distances are relatively short. In the Cosmological Realm, this time delay is crucial.

Second, in our house example, the distance between you and the house increased because you walked backward. In the Cosmological Realm, it is the expansion of space itself that causes distances to increase. This expansion isn't uniform like walking; the farther away an object is, the faster it moves away from you.

This means that by the time the light from the most distant galaxies reaches you, the galaxies have already moved far away.

When physicists use the apparent width of an object like a galaxy (or the average distances between galaxies) to determine the distance to that object, space's expansion and the relatively "slow" speed of light on cosmological scales have to be considered. Otherwise, the measurer will be tricked into thinking galaxies are where they appear. But they never are. Never.

More peculiar than this expansion of space-time and the attendant stretching of light is the impact that expanding space-time has on the energy content of light and matter. We now know that both lose energy due to the expansion of space-time, but in different ways.

In the Cosmological Realm, both the processes and the rules governing these processes diverge sharply from what we know in the Middle Realm. In the latter, matter rarely spontaneously gains potential energy, and the energy in light constantly increases. In the Cosmological Realm, it's the opposite. The expansion of space increases matter's gravitational potential energy while diluting its density. Light gets systematically redshifted— its wavelength stretched, its energy drained. Matter, too, loses energy: Thermal and kinetic motion fade as the universe cools.

In the end, everything winds down. If the expansion continues unchecked, the universe will reach a state of "heat death," a cold, dark equilibrium where no useful energy remains. No stars. No visible light. No life.

Luckily for us, that moment won't come for a very, very, very long time.

■ ■ ■

Now that you've seen how space-time behaves on the largest scales—shaping the expansion of the entire universe—it's time to zoom in. Let's look at how space-time behaves locally, in regions where matter clumps together: galaxies, clusters, and the cosmic web. These smaller-scale structures are sculpted by the same space-time, just playing by different rules in different neighborhoods.

Just as we can think of the living cell as the basic building block of multicellular life, we can think of the galaxy as the basic building block of matter in the Cosmological Realm. And like living cells, galaxies are as much a process as they are a thing. They come in different types, have internal structures, and interact to form cosmological superstructures.

Galaxies generally appear in three primary forms: small blobs, large disks, and massive spheres.

The small blobs, known as irregular or dwarf galaxies, lack a defined structure. They're simply clusters of stars and gas with no distinguishing shape, giving them a "blob-like" appearance.

When most people picture a galaxy, they often envision a disk galaxy. These typically feature a central red sphere of stars surrounded by a broad, flat disk of stars, dust, and gas, speckled with crimson, white, and blue. Disk galaxies are highly evolved structures, forming as smaller galaxies and gas clouds collide and exchange momentum. Through this process, the system stabilizes into a state that minimizes collisions, ultimately creating a disk shape.

Then there are the giant spheres or elliptical galaxies. In the universe's earlier days, when galaxies were much closer together, collisions were frequent. The telltale sign of past collisions is a spherical arrangement of stars. (Remember, this is what will happen after the Milky Way and Andromeda merge into the Hakeem Galaxy; it'll become one big sexy elliptical galaxy.) While disks emerge from an orderly evolution, collisions disrupt this order, scattering stars in random orbits. This explains why stars in the central bulges of disk galaxies are often arranged spherically. When disk galaxies collide, they form a new structure, often an elliptical galaxy—a massive, rounded galaxy with little gas or dust.

During such mergers, giant molecular clouds, which span hundreds of light-years, collide, while stars, tiny in comparison, rarely do. The collisions among GMCs set off bursts of star formation, using up most of the gas and dust while powerful stellar winds clear away the rest. Within about ten million years, the most massive of these newborn stars explode as supernovas, pushing even more matter out of the galaxy. An elliptical galaxy remains.

The regions of space where galaxies are packed closer together than usual are called galaxy clusters. Imagine looking at a deep-space image, such as those captured by the James Webb Space Telescope (JWST), of a massive galaxy cluster, like the Perseus cluster or the Coma cluster. At the cluster's heart, you might notice one or two gigantic elliptical galaxies dominating the view. These colossal structures often appear more than ten times the size of surrounding galaxies. They may even have multiple glowing cores, each the remnant of a previous galaxy that merged into the larger one.

Even more striking is the immense pool of hot gas spread throughout the cluster's interior, glowing faintly in X-ray images. When galaxies in a cluster collide, they release gas and dust, which heats to more than a million degrees Fahrenheit and fills the space between galaxies with plasma. This is known as the intracluster medium. Picture it like a cosmic soup, filling the gaps between galaxies with gas hot enough to emit X-rays. As the galaxies move through this hot plasma, friction strips away their gas and dust, adding even more material to the million-degree intracluster medium. Remarkably, the mass of this hot, diffuse gas is usually twice the total mass of the galaxies in the cluster!

At the opposite end of galaxy evolution are baby galaxies, the building blocks from which galaxies grow. These small galaxies, known as dwarf galaxies, are scattered throughout the universe and often look like faint, shapeless clouds of stars. Unlike the defined disks of spiral galaxies, such as Andromeda and our own Milky Way, or the smooth, oval appearance of giant ellipticals, dwarf galaxies are irregular and less structured. They hold fewer stars—often less than 1 percent of the Milky Way's stellar population—but they contain plentiful gas, fueling future star formation.

Dwarf irregulars are crucial in galaxy formation, much as planetesimals are in planet formation; they are the small pieces that collide and combine to create larger structures. Over time, these dwarfs grow and evolve: They might merge to form spiral galaxies and later develop into ellipticals through further collisions and mergers. This cycle illustrates the ongoing transformation within the Cosmological Realm, where galaxies and clusters follow a fundamental process of evolution.

Just as the Realm of Life and the Middle Realm are governed by the principle of least energy, galaxies in the Cosmological Realm follow a similar guiding principle: *zero and infinite curvature*. This principle reflects how space-time, shaped by the vastness and the intense gravitational forces at play, orchestrates the cosmic evolution of galaxies.

Stick with me here.

Water on Earth flows and collects in reservoirs based on the local topology of the land. It moves from higher elevations like mountains and hills in channels of increasing size—streams, creeks, rivers—and collects into depressions of increasing size: puddles, ponds, lakes, seas, and oceans. Water also cycles into and out of these systems through evaporation and rain, forming temporary reservoirs like clouds and fog.

Likewise, matter in the universe, primarily hydrogen gas, flows based on the topology of space-time. Instead of elevation as on Earth, space-time's topology is defined by the local curvature, determined by the density of mass and energy in that region. Our universe's voids are the emptiest regions that exist (and our universe is roughly 80 percent void) and are spattered with thin clouds of hydrogen gas. They are analogous to Earth's tallest and steepest mountains. Instead of water flowing down slopes to lower elevations, the matter in voids flows toward regions of greater space-time curvature. There, the matter pools into galaxies and galaxy clusters just as water does at the bottom of a waterfall.

Where there are galaxies, the mass-energy density is much higher, and space is much more curved, forming a gravitational depression similar to Earth's oceans or lake basins. Because there is so much local space-time curvature around galaxies, space does not expand here. The most extreme gravitational depressions are caused by galaxy clusters, which have masses one thousand to ten thousand times that of the Milky Way, and galaxy superclusters, which are at least ten times more massive still.

Our analogy with Earth's water cycle can take us only so far for one simple reason. Earth's surface is two-dimensional, whereas space is a three-dimensional volume. The universe's galaxy clusters are like the nodes in a 3D network connected by cooler filaments of hydrogen plasma and galaxies. We refer to the filament plasma as the warm-hot intergalactic medium, or WHIM. It's thought to contain 40–50 percent of all the universe's ordinary matter. Galaxies outside clusters are typically embedded in the WHIM, as is the case for the Milky Way.

Plasma in the clusters, exploding stars, and jets from black holes are like evaporation for galaxies. They strip and explosively eject matter from the galaxies into the WHIM. This ejected matter will ultimately rain back onto the galaxy or some other galaxy. Likewise, hydrogen rains from the voids onto the filaments and galaxies.

We don't know precisely how this universal system of flows and reservoirs evolved from the earliest times. Although, as I will discuss in the next chapter, we have identified the process determining where galaxy clusters and galaxies form in our universe. What we have gotten wrong is *when* galaxies formed in our universe. Images from the James Webb Space Telescope have shown that evolved disk-shaped galaxies existed much earlier in our universe than can be accounted for by our model of merging dwarf irregular galaxies.

Regardless, the general trend in our universe's curvature is evident. The space in our universe's large voids, measuring 150 million light-years across on average, constantly expands, evolving toward global emptiness and zero curvature. Conversely, where mass and energy concentrate in galaxy clusters and in the WHIM, our universe evolves toward increased local curvature. Ultimately, local curvature approaches infinity inside the densest mass concentrations of all: black holes. As the energy density goes to infinity near a black hole's center, so does space-time curvature. Space gets compressed as close to zero as possible, and time, with its opposite sign, is stretched to infinity.

We started our journey through the Cosmological Realm by looking at space-time on the largest of scales. We learned how space-time dictates the motions of all objects in the Cosmological Realm. Then we moved to the local scales of galaxies and clusters. However, one thing I haven't shared with you is that by the opening of the twenty-first century, we had realized a considerable discrepancy between what our physics predicted about our universe's expansion and the motions of galaxies and what we actually observed them to be.

Short story: There isn't enough energy in the observable universe to account for how galaxies move in the Cosmological Realm. And the discrepancy isn't tiny; it's three or four times too small. By some reckonings, it's *twenty times* too small! When we peer out and add up all the energy in our universe, we come up too short. All this matter should never have been able to come together. Galaxies should not exist. Galaxy clusters should not exist.

And life? That should be a nonstarter.

DARK REALM

Outside of a dog, a book is man's best friend. Inside of a dog, it's too dark to read.

—Groucho Marx

Life is only able to exist because something ain't right in the Cosmological Realm. On scales at and beyond a galactic arm, beginning at around twenty thousand light-years, our universe reveals a deep and mysterious inconsistency. This puzzle defies everything we thought we knew about the stuff that makes up, well, reality.

Our most trusted and rigorously tested scientific theories fail us on these massive scales. Put simply, nothing here appears to move according to the known laws of physics. Stars and galaxies behave as if unseen forces are at play, guiding their motions and shaping their structures in ways we cannot fully explain. Space-time at cosmological scales appears to brim with ghostly, unseen matter possessing an immense collective mass that binds the universe together.

At the same time, an even more mysterious source of energy imbues the universe's giant voids, producing an enigmatic force that pushes galaxies apart and accelerates the expansion of space-time.

These mysterious discrepancies have led us to propose two of the most astonishing ideas in modern physics: dark matter and dark energy. Together, they form the invisible mirror side of our cosmos, accounting for more than 95 percent of our universe's mass-energy density. Yes, 95 percent. While dark matter pulls and binds, creating the cosmic web that is home to the warm-hot intergalactic medium, galaxy clusters, and galaxies themselves, dark energy pushes, expanding the fabric of space-time and seal- ing the universe's ultimate fate.

But what exactly are these "dark" phenomena? Are they real, or are they mathematical illusions conjured up to span gaps in our knowledge? Most crucially, could our very existence be tied to the mysteries of this Dark Realm?

If we can better understand all that is dark, can we better un- derstand all that is light?

▪ ▪ ▪

All right—let's get our Dark on.

First, how can we even say that there must be some kind of "dark" stuff out there undergirding the entire cosmos if we can't see any of it?

Well, consider space-time, another thing we can't see, but we know it is there. Recall from the previous chapter that local space-time curvature is determined by mass-energy densities— planets, stars, GMCs, galaxies—and the paths of matter and light are tied to this space-time curvature. This reality was summed up nicely by the physicist John Wheeler in 1990: "Spacetime tells matter how to move. Matter tells spacetime how to curve."

Even though we can't see the substance of space-time, we can see its effects. It is now routine for physicists to "see" space-time curvature by observing the movement of matter and light near and within galaxies and galaxy clusters. Calculating the curva- ture provides insight into the mass-energy density of any given region of space. In short, the motions of matter and light reveal the space-time curvature, which allows the *total* local mass-energy density to be "seen."

We physicists have repeatedly measured space-time curvature across many terrestrial and celestial environments. Everything works as predicted when these measurements are performed in the Middle Realm. Apples fall, jumping toddlers return to Earth, planets orbit. But in the Cosmological Realm, we find that the local space-time curvature around galaxies and galaxy clusters is far *greater* than what can be accounted for by the mass-energy densities we observe in the emitted light.

As I said, something ain't right in the Cosmological Realm. These deviant observations suggest that either a significant amount of "stuff" is being concealed from us or we have a limited understanding of how gravity and space-time curvature operate.

Here's how the famed astronomer Vera Rubin conclusively illustrated anomalies like these back in 1970.

From our perspective, observing a disk-shaped spiral galaxy edge-on is like looking at a pancake from the side. In these cases, the galaxy's rotational motion becomes discernible from the light its stars and bright nebulae emit as one side of the edge-on galaxy moves toward us while the other moves away. When objects move away from us, their light experiences a redshift, resulting in longer wavelengths. Conversely, objects moving toward us provide light with a blueshift, leading to shorter wavelengths. The apparent shift of wavelengths caused by relative motions is known as the Doppler effect.*

If you live in a city, you might have noticed the Doppler effect whenever an ambulance screams toward you with its siren blar-

---

* This is totally different from cosmological redshifts. These are a *real* change in light's wavelength—light gets longer—not an *apparent* shift in light's wavelength due to relative motion.

ing. As the ambulance gets closer, the sound seems to "bunch up," resulting in quicker high and low pitches in the wail. This phenomenon is known as blueshift. However, once the ambulance passes, the sound source moves away, causing the sound to "stretch out," leading to slower high and low pitches. This effect is known as redshift. These changes in pitch and wavelength occur due to relative motion, even though the siren itself emits a constant sound. The Doppler shift, which affects all waves, whether they're present in light, sound, water, or something else, is our main tool for measuring astronomical motions. Furthermore, the magnitudes of these wavelength shifts relate to speed; in this case, the speed of an edge-on spiral galaxy's rotation.

According to the well-tested laws of physics, objects farther from a galaxy's center of mass should orbit the galaxy more slowly due to less curved space-time. However, this is not the case! Objects in galaxies move at a similar speed from near the center all the way out to the visible edge. Since space-time curvature dictates this motion and is directly related to mass-density concentrations, we can infer that the entire galaxy we are observing must lie at the center of a much larger and invisible—or "dark"—mass concentration.

How can this be?

Let's consider a yo-yo. If you hold a yo-yo's string and begin to whip the yo-yo around in a circle, the string tightens. The force the string exerts on the yo-yo acts along the string and points toward your hand. We call this force tension. It is entirely analogous to the gravitational force that stars "feel" as they orbit the supermassive black hole at the center of their galaxy. (Our earlier yo-yo example showed how an object can constantly accelerate in a direction without obviously moving in that direction.)

Okay, stop spinning the yo-yo and let it come to rest. When the yo-yo hangs straight down at the end of the string, the string's tension equals the weight of the yo-yo, and the direction of this tension still points up along the string to your hand. Start spinning it again. Get it going good and fast, so fast that you can move your hand above your head and spin the yo-yo in such a way that the hand-string-yo-yo path is now parallel to the ground (something you could never do if the yo-yo were not spinning—not on Earth anyway). From this point onward, rotating your hand more quickly does not change the path of the yo-yo, but it does increase the tension in the string. The faster you rotate your hand, the greater the tension. If the rotational speed becomes high enough, the tension may exceed the string's strength, causing it to snap and allowing the yo-yo to escape.

This example illustrates the relationship between a yo-yo's rotational speed and the tension force that keeps it confined to a circular motion. We can calculate the tension force by measuring the yo-yo's rotational speed. This principle also applies to stars and nebulae orbiting a galaxy: A faster rotation around a galaxy's center necessitates a more powerful force to keep stars in their observed orbits, and measuring the rotational velocity allows us to determine the confining force.

But there's more! We physicists are quite clever at eking out physical information about what we can't see from the little that we can. If you imagine a series of yo-yos along our string, like groups of hot, bright stars and giant glowing nebulae, we can measure the rotational velocities of each from the galaxy's center to its outermost edges. Then we can plot the galaxy's *rotation curve*.

From this curve, we can calculate the galaxy's mass versus distance from its center. When Vera Rubin calculated the rotation curve of the Andromeda galaxy in 1970 from her Doppler measurements, she found that the farther out she looked, the more mass the galaxy appeared to have. When all the mass was added

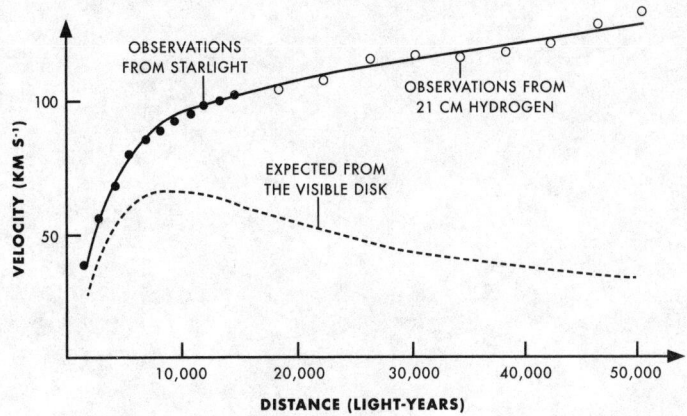

Fig. 12: Spiral Galaxy Rotation Curves
Observed rotation curve of a spiral galaxy vs. expected curve based on visible matter—evidence for dark matter.

up, she found the galaxy contained twelve times more "invisible mass" than visible mass.

We consistently observe this phenomenon across the thousands of galaxies for which we have measured detailed rotation curves. By comparing the "illumination mass" of any given galaxy (the mass derived from the light emitted by a galaxy's visible matter) with its dynamical mass (the mass derived from the movements of the galaxy's visible matter), we find that visible matter makes up only about 10–20 percent of the total mass required to explain the observed rotation curves. This means that a significant amount of "dark matter" accounts for the remaining 80–90 percent of any given galaxy's mass; this is the source of all that extra space-time curvature.

Extra space-time curvature reveals itself at larger scales through similar oddities in galaxy clusters. While the ultra-massive elliptical galaxies near the center of galaxy clusters are virtually at rest

FIG. 13: COMA GALAXY CLUSTER
A dense cluster of galaxies whose member motions and gravitational lensing helped reveal the presence of unseen mass.

relative to space's expansion, the surrounding constituent cluster galaxies rapidly scurry about within the cluster.

We can use the same technique of comparing the cluster's illumination mass with its dynamical mass. We get the first by measuring the light emissions from all the galaxies and plasmas that make up the cluster (the cluster's illumination mass). Next, we can use the Doppler effect to measure the speeds of galaxies as they dash about within the cluster. As with individual spiral galaxies, we invariably find that within clusters the illumination mass accounts for only 10–20 percent of the total mass required to explain the observed motions. The remaining 80–90 percent of the cluster's mass belongs to "dark matter," which is otherwise undetectable save for its gravitational effects.

Let's imagine another mental model along the lines of the yo-yo to help explain the extra curvature found within and around clusters. Here, we'll delve into the concept of gravity wells. A gravity well is a funnel-shaped model representing the gravitational field surrounding a massive object, such as a star, galaxy, or black hole. This concept helps visualize why objects move as they do under gravity, resembling sliding down a slope toward the center of mass. The most important thing to remember here is that the more massive and dense an object is, the deeper and steeper its gravity well. (Black holes, which are the densest structures we're aware of, have gravity wells that are effectively infinite in depth.)

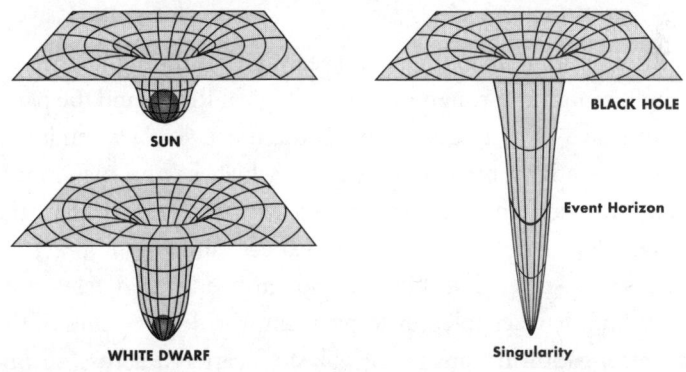

Fig. 14: Black Hole Space-Time Curvature
The warping of space-time caused by a black hole. The steep gravity well explains the inescapability of the event horizon.

Gravity wells help describe gravitational potential energy, escape velocity, and the curvature of space-time. Escape velocity—the minimum velocity required for an object to escape a gravitational pull—is related to the depth of the object's gravity well. The deeper the gravity well, the more pronounced the curvature of space-time, and the more energy an object requires to escape it. Consequently, a higher escape velocity is needed to break free. For instance, the escape velocity from Earth is approximately 11.2 kilometers per second, while the escape velocity from the Sun is around 617.5 kilometers per second. From the Moon, it's a paltry 2.4 kilometers per second.

Analyzing a cluster's total mass distribution makes calculating its gravity well and escape velocity possible. Surprisingly, the galaxies swimming around inside clusters seem to be moving much faster than the clusters' escape velocity, suggesting that the clusters should not exist; the minor galaxies should "escape" and be done with the cluster. Yet these clusters *do* exist, and these minor galaxies cannot escape. The gravity well in which the galaxies are moving is deeper than can be accounted for by the cluster's total illumination mass.

Another way this extra mass reveals itself is through our study of light's motion through the Cosmological Realm and the paths it must take as it streams through the universe. As a reminder, general relativity predicts that light, which has no mass, must follow the contours of space-time's curvature rather than the straight-line paths we intuitively expect, and this is precisely what we observe. The biggest and most advanced telescopes we've built have enabled us to peer into the deep reaches of the universe, capturing images of galaxies scattered across an immense, dark canvas. When we observe a cluster of galaxies, we often see more galaxies lying far beyond, serendipitously aligned along the same line of sight. The journey of light from these distant galaxies is anything but straight; it is bent and redirected as it traverses the curved fabric of space-time within and surrounding the cluster. This bending of light is known as gravitational lensing, a spectacular display of space-time's influence on the path light must take.

Similar to how a well-crafted magnifying glass focuses light and sharpens an image, a galaxy can bend light nearly symmetrically around itself, making background objects it passes in front of appear brighter. The most spectacular example of this effect is an Einstein ring.

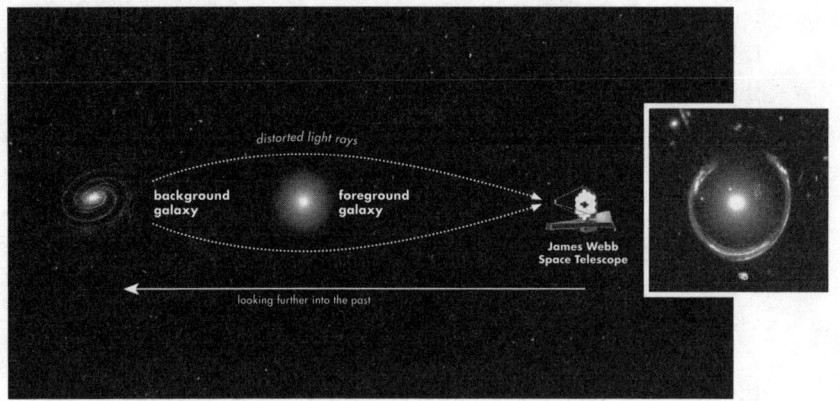

Fig. 15: Einstein Ring
An extreme example of gravitational lensing where light from a background object is bent into a near-perfect circle.

It happens when light from a distant object, like a galaxy or the hot disk of plasma surrounding a supermassive black hole, is bent by the gravitational influence of a massive object positioned directly between the light source and the observer.

For a complete Einstein ring to form, several conditions need to come together. The alignment between the light source, the gravitational lens, and the observer must be nearly perfect to ensure that the gravitational lens bends the light symmetrically around the intervening object, creating the appearance of a ring.

The object acting as the gravitational lens must have sufficient mass to create a significant curvature in space-time, capable of noticeably bending light. Only something truly enormous, like a galaxy, a galaxy cluster, or a supermassive black hole, packs the gravitational heft to bend space-time so acutely.

Additionally, the source of light must be located far behind the lens. The distance between the observer, the lens, and the source determines the apparent size of the Einstein ring, which depends on the lens's mass and the three distances involved: the distance to the lens, the distance to the more distant object being lensed, and the distance between the two objects. You would think that measuring the first two distances would give you the third by calculating their difference. But the Cosmological Realm is sufficiently weird that the distance between the objects is not the same as the distance obtained by calculating this difference. (And of course, "distance between objects" is also not an accurate description, since the object that emitted the light and the object that lensed it are not in the same locations once the light hits our telescopes.)

Of course, gravitational lenses are not optical lenses, and of the hundreds discovered so far, only a few are so perfectly tuned to generate Einstein rings. Most often, gravitational lenses warp images in the same way a fun-house mirror does. Individual galaxies in the distant background of space can be stretched into distorted arcs and rendered into multiple "copies" of the same galaxy, which appear symmetrically arranged around the intervening gravitational lens.

The variations in the type and strength of gravitational lenses enable us to map their gravity wells and conduct some remarkable additional science. Take lensed exploding stars. Imagine a distant galaxy hosting a supernova, where the light from the explosion doesn't travel straight to us. Instead, it gets bent and split by the space-time curvature variations of a galaxy cluster lying between us and the supernova. This bending creates multiple images of the supernova, each appearing at a slightly different position, and, crucially, at slightly different times.

The reason for the time differences can be attributed to two primary effects. First, the paths that the light from the supernova takes to reach us are not all the same length. Some are more direct, while others are more circuitous. Second, these light paths pass through regions of space-time distorted by the lensing cluster's gravity, which slows down the light's travel, almost like cosmic molasses. Yeah, you heard that right. The speed of light is slower in regions with greater space-time curvature as a consequence of time dilation.

Together, these factors create a delay. The same supernova explosion can appear at different times and locations around the lens, much like a rerun of the same cosmic fireworks show. In 2021, astrophysicists reported a lensed supernova with a *twenty-year* time delay.

By carefully measuring these time delays, we can infer a lot about the cluster and the observable universe as a whole. For starters, the delays depend on how light travels through the expanding fabric of space-time, which determines the current rate of universal expansion.

The time delays also encode information about the lensing cluster itself. By analyzing the delays and the lens's gravitational influence, we can map the distribution of the cluster's mass, including its dark matter. It's like having a flashlight that reveals not just what's visible but also the invisible scaffolding of the cluster.

"Strong" and "weak" lensing are used to create detailed maps of dark matter distributions around and within galaxy clusters. It's like drawing a topographic map of a landscape, where the peaks and valleys represent regions of stronger or weaker gravity, which simply denote more or less space-time curvature. From these maps, we can then calculate the relative depth of gravity wells. Again, we have done this, and they are always way deeper than what we would expect from the visible matter alone.

What does all this tell us? When we look out at the cosmos and try to account for all the matter that should create such deep gravity wells and excessive space-time curvature, we come up short. There isn't nearly enough visible "stuff" to explain the gravitational effects we observe. Most astronomers have concluded that an invisible mass-energy density surrounds almost all of the universe's galaxies and clusters. This is dark matter.

We call it dark because it neither emits, nor absorbs, nor reflects light. However, it's not really dark. A lump of coal is dark, and if I throw it at you, it would rebound off your forehead, and you'd feel it.

In contrast, you've had copious amounts of dark matter passing through your body every second of your life, and you've never noticed. You can't see it or feel it. And even though it does affect great concentrations of matter like galaxies and galaxy clusters, it ignores smaller things like stars and planets—and you. We mean less to it than it does to us. Heck, it doesn't even interact with itself! As far as we know, dark matter *only* interacts with space-time. For these reasons, I prefer to refer to it as phantom matter.

What might this phantom matter be? Frankly, no one knows. As I confessed earlier, it's one of the universe's most impenetrable unsolved puzzles. We only know it's there because of how it affects the things we can see, much like how you can tell the wind is blowing by watching tree branches sway, even if you can't see the air itself.

We do know some things about dark matter by understanding what it *isn't*. First, it's not made of the same kinds of atoms that build up our everyday world—the stuff that makes up everything from mountains and oceans to the very cells in our bodies. This ordinary matter, or baryonic matter, interacts with light. We also know it's not made up of gas clouds, free planets, or dim objects like burned-out stars. Those were once considered possible culprits for all the universe's missing mass, but they don't come close to accounting for the gravitational effects we observe on cosmic scales.

So, what *could* this elusive substance be?

We have a few leading theories. For most of the last four decades, since we've been trying to detect it directly, the primary candidate has been something called a weakly interacting massive particle (WIMP). Imagine a ghostly heavyweight that can pass through walls without causing even the slightest vibration. WIMPs would be massive enough to exert gravity but so shy about interacting with ordinary matter that they're almost impossible to catch in action.

The best idea for what the WIMPs are is something called supersymmetric particles, a consequence of string theory. String theory was once considered humanity's best idea for creating a unified system that explained the universe at every scale—the long-sought "theory of everything"—but has since fallen out of favor.

Coincidentally, my first physics research gig was on a team trying to detect WIMPs. That experiment, conducted in the basement of UC Berkeley, eventually evolved into the Cryogenic Dark Matter Search. Since then, scientists have developed increasingly sensitive detectors and fired particles at each other in massive colliders, such as the Large Hadron Collider in Geneva, Switzerland, all in the hope of catching a glimpse of WIMPs. But after decades of searching, scientists have turned up nothing. The detectors continue to come up empty, and the particle collisions have failed to reveal anything new. Most physicists now believe WIMPs, and the string theories that predicted them, probably aren't the answer.

So, dark matter remains a shadowy companion to everything we know, a silent architect shaping the cosmos while keeping its true identity firmly hidden. What's important to understand for our purposes is its role in generating our life-sustaining universe. For without this phantom dark matter, we would not exist.

▪ ▪ ▪

So now we know that every non-dwarf galaxy and galaxy cluster is embedded in a giant halo of dark matter that provides the gravitational stability necessary to tame fast-moving stars within galaxies, as well as fast-moving galaxies within clusters. In our early universe, dark matter also provided the gravity wells that would later determine where galaxies formed. We have confirmed this by studying the oldest light in the universe.

Our universe is filled with light from its earliest times, but that did not originate from stars or galaxies. Today, this light is at very low energies, and it is not visible. But there's so much of it that its total energy density swamps the energy density of all the other light in our universe. The photons associated with this low-energy light outnumber protons by a factor of a billion to one! Because this light falls within the microwave region of the electromagnetic spectrum, we refer to it as the cosmic microwave background radiation, or CMB.

Recall that the longer the light's wavelength, the lower its energy. This means that back when this light began its journey to us, about five minutes after the universe sprang into being, it was one hundred million times *more* energetic than it is now. For the first fifty thousand years of our universe's existence, the energy density from this light was totally dominant. It alone determined the global curvature of space-time. All other energy densities were negligible in comparison.

But here, we are concerned with how dark matter and light worked together in the first few hundred thousand years of the universe's existence to determine where galaxies and galaxy clusters would form. Dark matter created large gravity wells for matter to fall into. Light provided a reverse pressure that counteracted this process, eventually forcing matter back out of the gravity wells due to its own pressure.

You may wonder how light could exert enough pressure on a galaxy cluster's mass of matter falling into a giant gravity well to push the matter back out. We're talking about masses up to one hundred trillion times that of the Sun after all. Forgive me, but that's nuts!

The light accomplished this impossible-sounding feat because matter was too hot in the early universe to form atoms. It was in a hot plasma state, and the electrons couldn't go about the job of capturing protons. Not incidentally, free charged particles inter-act incredibly strongly with light. This is the reason that metals are shiny. Their surfaces are full of free electrons. When you look at metals, the shine you see is light that has been absorbed and re-emitted by these free surface electrons. In our early universe, this strong interaction caused light to influence matter's motions.

But do you know what was *not* affected by this light? Dark matter.

Look at the four ingredients of our early universe: baryonic matter (at this stage, a plasma of free charged particles), dark matter, light, and space-time. Matter and light can't be respon-sible for the formation of early galaxies. Light moves at the speed of light, so it can't stop moving to create the local space-time curvature needed for seeding galaxies. And since the plasma-state matter was affected by all this fast-moving light applying immense pressures on it, the matter couldn't successfully clump together, either. Dark matter, on the other hand, does not inter-act with light, matter, or itself. It interacts only with space-time.

The issue with galaxy formation in the early universe is that the CMB clearly shows us that baryonic matter in our early uni-verse was distributed uniformly to a very high precision. Put simply, galaxies, which are variations in the universe's material, should not have formed. The matter we see should have been spread out and never come together. But galaxies *did* form. This occurred solely due to a peculiar property of the Quantum Realm, which made the universe's dark matter ever so slightly nonuniform. This nonuniformity at the subatomic level resulted in initial differences in dark matter distributions, which in turn led to the genesis of galaxies—and eventually to us.

As I've written before: Wild!

•••

For nearly four hundred thousand years after the Big Bang, or-dinary matter and light were engaged in a tug-of-war. Meanwhile, dark matter followed the curvature of space-time, flowing into gravity wells of its own creation. As these wells grew deeper, baryonic matter followed, pouring into them like water seeking lower elevations. Gravity amplified these over-dense regions, pulling more matter into them, creating compression.

Once the matter became dense enough, the intense photon pressure of the tightly coupled light and baryons counteracted this compression, leading to an outward push. This interaction between gravitational attraction and photon pressure set up oscillations, analogous to sound waves propagating through the plasma. These oscillations created spherical waves of compression (where gravity dominated) and rarefaction (where photon pressure rebounded baryonic matter outward), with their frequency determined by the speed of sound in the plasma. We call the imprint of these waves on the CMB acoustic oscillations.

These oscillations continued until our universe cooled enough for protons and electrons to bond into neutral hydrogen during a period known as recombination. During recombination, our elementary friends, the electrons, got busy hooking up with all those protons out there. Baryonic matter went from being an opaque plasma to a transparent gas. Photons decoupled from matter and escaped, forming the CMB and thereby freezing the acoustic oscillations in place.

The result was a characteristic pattern of temperature variations imprinted on the CMB. Regions that were in compression now appear colder in the CMB fluctuations, while those in rarefaction appear slighty hotter. This is because where there was more matter, photons lost energy climbing out of those gravity wells. The patterns provide a snapshot of the universe at a time when it was just 380,000 years old, offering a window into the physical processes and conditions of the early universe.

Amazingly, we can predict the sizes of the largest oscillating sites and how large they should appear in our sky today. Remember in the Cosmological Realm where I told you that our universe's global curvature has been measured to be nearly zero, which is to say that ours is a flat universe? We determined that by comparing the sizes of CMB temperature variations with the size we predict they should be.

The angular size of the temperature variations in the CMB caused by acoustic oscillations, as measured by the Planck satellite, is approximately one degree, just as predicted. Significant deviations from flatness would distort this angular size either smaller (for positive curvature) or larger (for negative curvature). But the Planck satellite's results show no such distortion within the precision of the data. This means that the global space-time curvature of our universe is remarkably close to flat.

Similar to the signal in CMB variations, baryon acoustic oscillations are the imprints of sound waves on the *arrangement* of galaxies in the universe. When photons decoupled from the baryons, the oscillations were also frozen into the distribution of matter. They established a characteristic scale equal to the size of the universe's sound horizon at the time, corresponding to the maximum distance these waves could travel before decoupling. This characteristic scale is imprinted on the arrangement of the galaxies, manifesting as a slight excess in the number of galaxy pairs separated by the sound horizon distance, about 150 megaparsecs today.

When we view galaxies, we find that they're all embedded in a giant halo of dark matter ten times the size of the galaxy. But as we know, dark matter cannot cool by emitting light, so it cannot lose energy and condense into structures like stars or planets. Dark matter halos remain bloated and fluffy. As a result, the Dark Realm has no known structures at the scale of the Middle Realm or smaller. Its only arrangements are on cosmological scales.

As electrons bring protons and other nuclei together to form atoms, molecules, and larger matter concentrations like stars, planets, and life in the Middle Realm, dark matter brings this baryonic matter together in the Cosmological Realm. And by doing so, it performs a vital role in our existence. Dark matter provides the gravitational stability that allows galaxies, clusters, and the WHIM to exist. And now, after around fourteen billion years, here you are, reading this book.

Wild.

■ ■ ■

So, dark matter is the prime player in binding the regular matter of our universe together. Dark energy, on the other hand, does something different: It pushes our universe apart. And while dark matter reveals its presence on the scales of galaxies and galaxy clusters, dark energy reveals its existence on even larger scales.

The most enormous structures in our universe are WHIM galaxy filaments, which appear like the colossal threads of some cosmic three-dimensional spiderweb. These filaments are peppered with myriad galaxies, tracing out the invisible lines of dark matter we physicists call the cosmic web. Picture a universe-sized web, with galaxies illuminated by the glow of ancient starlight,

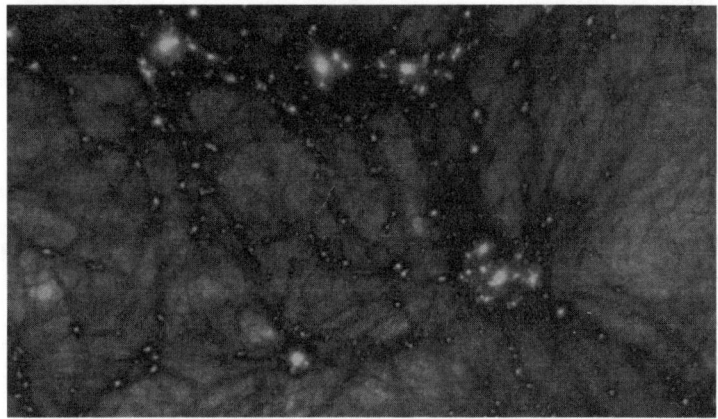

FIG. 16: COSMIC WEB OF GALAXIES ON DARK MATTER FILAMENTS
The large-scale structure of the universe: galaxies clumped along dark matter scaffolding spanning billions of light-years.

sparkling like drops of water caught in its threads, stretching out as far as we can see and beyond.

In between these filaments lie the cosmic voids—yawning, pitch-black chasms that stand in a kind of negative-space relief to the cosmic web. These are cosmic deserts, stretching tens of millions to hundreds of millions of light-years across. Here is where space itself breathes and grows. It is within these vast, haunting voids that the universe stretches its fabric most profoundly, pushing the cosmic web and its attendant galaxies ever farther apart.

And as you might expect, something at these scales also ain't right. The expansion rate in the voids—that Hubble constant we've touched on—does not align with our current laws of physics.

We can't look to yo-yos to explain this discrepancy, so let's go scuba diving instead.

Imagine you're a diver, and you're cruising around a vast, fluid-filled tank—like the Super-Kamiokande neutrino detector, buried one kilometer beneath Mount Ikeno in Gifu Prefecture, Japan. As you swim around this 50,220-cubic-meter enclosure, picture an intricate system of pipes stretching through the tank like cosmic filaments, intersecting at lumpy, quasi-spherical nodes. Tiny holes pepper these pipes and nodes, acting like drains that draw in water, each one a swirling miniature whirlpool.

Between these pipes and their drains are broad stretches of open water: our analogue for the giant cosmic voids stretching across our universe. These silent spaces are like calm, untouched water. They're not totally pure—the voids do have some stuff in them, just very, very little—but are laced with dissolved minerals. The water is near freezing, and some minerals condense into ice grains that float around like suspended motes. These collect near the tiny drain holes, accumulating into delicate icy flakes that adhere to the pipes and nodes. In this analogy, the plumbing represents dark matter filaments, while the ice flakes are galaxies clinging to them.

But then a subtle ripple stirs around you as if the water were moving in all directions at once. Your depth gauge shows you plummeting, and you sense the pipes drifting away from you. The tank is expanding, yet somehow still full to the brim even though no new water is flowing in. It's as though every drop of water were expanding—except the water near the pipes. The open water stretches more and more rapidly, just like the universe itself.

There is virtually zero matter density in these voids bereft of dark matter, meaning there is practically no gravitational resistance. And absent this resistance, space is free to stretch. Here, the mysterious dark energy of the Dark Realm pushes ever outward with a quiet, relentless intensity, powering an invisible flow of space that accelerates the distances between galaxies and galaxy clusters. Bound to their cosmic dark matter filaments, the galaxies remain connected and intact while the fabric of space in the voids expands away from them.

The 1998 discovery that the universe's expansion is accelerating under the influence of the mysterious dark energy was perhaps the most unexpected cosmological finding since Edwin Hubble uncovered the universe's expansion seventy years earlier. Prior to this discovery, we presumed that the mutual gravitational attraction between galaxies would cause the universe's expansion rate to slow with time. Here's an example to illustrate why.

Imagine throwing a baseball straight up from Earth's surface. It has three possible fates depending on the speed at which it leaves your hand. The first is that the ball rises, stops, and then falls back to the ground. The second occurs if you can, somehow throw the ball upward at a speed exceeding Earth's escape velocity of around twenty-five thousand miles per hour. Sure, the Incredible Hulk, LeBron, and I can do that. And if you believe strongly enough, maybe you can, too (no, you can't). But if you did (you won't), the ball would travel upward forever and never return to Earth.

The third option is throwing the ball at just the right speed so it reaches a near standstill relative to Earth's surface. It will perpetually move away, but at a gradually decreasing pace. This is akin to trying to walk through a doorway by taking half the distance to the door with each step. You will always be moving toward the door, but you'll never manage to pass through it. Eventually, it will appear as if you were motionless, even though you're always slowly approaching the door. (For my philosophers, this is known as Zeno's paradox of motion, and at its most extreme it says that all motion is but an illusion! But I digress.)

All three outcomes contend with a common factor: The baseball's speed will decrease once it leaves your hand. This is because the ball must overcome Earth's gravitational pull, which requires it to expend kinetic energy. This is our everyday experience in the Middle Realm.

But what if the ball didn't slow down as it rose, but instead slowed down briefly before accelerating upward faster? It would be an incredible experience! The ball would gain kinetic energy instead of consuming it to reach higher heights. This only happens in the Middle Realm if there is an energy source, such as a rocket engine or a stellar wind, to accelerate the object. Well, our universe did not get the memo. Its expansion rate behaved just as if all the galaxies had been thrown "upward" from each other, gravitationally slowed down for a time, and then sped away again as if gaining energy!

Remember when I mentioned earlier that my first physics research experience was at UC Berkeley, back in 1991? One of the postdoctoral researchers I worked with that summer was Saul Perlmutter, the man who shared the 2011 Nobel Prize in Physics for discovering dark energy.

Perlmutter's team achieved this by developing groundbreaking methods to locate and measure a specific type of exploding star—a Type Ia supernova—at various distances. They collected these measurements in sufficient numbers and with enough precision to reveal the universe's changing expansion rate over time.

Here was the challenge. No one knows when and where a star

is going to explode in our universe. Saul invented a revolutionary method called image differencing that allowed for the efficient discovery of supernovas in digital images. He leveraged the power of digital detectors and advanced computational techniques to identify faint, transient astronomical objects against a crowded background of stars and galaxies—something that was nearly impossible to do previously using images on photographic film.

The process involved taking two images of the same patch of sky at different times. The first image, referred to as the reference image, served as a baseline, while the second, the science image, was captured at a later time. These images were carefully aligned and normalized to ensure direct comparison.

Perlmutter developed an algorithm that subtracted the reference image from the science image, effectively canceling out all objects that remained constant between the two exposures. This process left only the differences—namely, transient or variable objects, like newly appearing supernovas. These residual signals were then analyzed to distinguish genuine supernovas from artifacts caused by noise, image imperfections, or other transient phenomena, such as passing asteroids.

The innovation was not just in the differencing technique itself but in its efficiency and scalability. Perlmutter's method made it feasible to sift through vast amounts of digital data, identifying potential supernovas quickly and reliably. Once candidates were found, they were further examined to confirm their nature as Type Ia supernovas, which are characterized by a particular brightness profile and spectroscopic fingerprint.

What makes Type Ia supernovas unique is that their absolute brightnesses are related to the way they brighten and fade after exploding. This means we can calibrate them to determine their actual brightness. Once we compare their apparent brightness with their known brightness, we can calculate their distance. From this distance, we can calculate the amount of time their light has taken to travel to us.

To measure space-time's dynamics, we need one more observation from the supernova—its spectrum at peak brightness. From the spectrum, we can measure the cosmological redshift of the supernova. Recall that light gets stretched by the same factor that space expanded while the light traveled through it. These two observations, of the time since the explosion and of the amount our universe grew while the light was traveling to us, give us the size of the universe versus time!

A form of this plot is known as a Hubble diagram. Different universe expansion scenarios will have unique Hubble diagrams. The measured Hubble diagram for our universe shows that it expanded at a decreasing rate for its first nine billion years. But then, five billion years ago, something changed drastically. The density of matter dropped below the density of dark energy. At this point, dark energy became the dominant component of the universe's energy budget, leading to an accelerated expansion of the universe.

This shift marked a profound change in our universe's dynamics as gravity's attractive force, governed by matter, was increasingly overwhelmed by dark energy's repulsive effect, driving the accelerated expansion we observe today.

We refer to this energy as dark not because it's ghostly or ephemeral like dark matter but because we genuinely have no idea what it is—we're in the dark.

While dark energy and matter are separate phenomena and quite different in many ways, they have some intriguing similarities beyond only revealing their existence on cosmological scales.

The first resemblance is that dark matter and dark energy, together, comprise roughly 95 percent of the universe's total mass-energy content. Dark matter accounts for about 27 percent, while dark energy contributes two and a half times as much at 68 percent. This leaves only about 5 percent for baryonic matter (such as stars, clouds, plasmas, and galaxies) that we can directly observe. These strange, dark substances make up almost *everything* that exists.

We're the weird ones!

But that does not mean dark energy is not strange. One of the most difficult-to-understand facts about dark energy is that it possesses positive energy but negative pressure. So, instead of being an energy density that creates attractive gravity (like dark matter), dark energy creates *repulsive* gravity. If that sounds like a contradiction to you, you're not alone! Let's unpack this.

We experience both positive and negative pressures all the time, but we don't tend to use those terms. We refer to positive pressure simply as pressure, and negative pressure as tension. When you compress a spring, it creates positive pressure, which makes the spring want to expand back to its equilibrium length. Conversely, if you pull the spring from both ends, it becomes imbued with negative pressure, or tension, causing it to want to shrink back to its equilibrium length.

More to the point, fill a chamber with gas, and the gas molecules will apply outward pressure on the container's walls. If one of the chamber walls is allowed to move—let's say we affix it with an arm and turn it into a piston—then the pressure from the gas will push the piston out. We could utilize this motion to extract energy. And we do! This is how car engines work. An expanding gas of ignited gasoline vapor trapped inside an engine block pushes a team of pistons, which then turn a crankshaft, which moves through a series of gears and ultimately turns the car's wheels. You're on your way. Thanks, positive pressure!

But what if the chamber is filled with nothingness? No gas. Just space. In that case, you must do work to pull the piston out because you're fighting against a *vacuum*. Vacuums don't push; they suck. You have to expend energy to pull the piston out. Since energy is conserved in local systems like this, where did the energy go that you put into expanding the vacuum? It entered the new vacuum created in the chamber, resulting in an even stronger vacuum. If you want to keep pulling that piston, you're going to have to pull harder and harder, or punch a hole in the contraption that defeats the vacuum altogether.

Examples of this are not common, but they're not uncommon either. Perhaps you liked to take baths when you were young, and while in the bath maybe you liked to play with a plastic cup. You could collect water in the cup, pour it over your head or shoulders, push air bubbles down to the bottom of the tub and release them—and also trap water in a vacuum. If you submerged the cup and let it fill with water then turned it upside down so that its open end was pointed toward the bottom of the tub, you could then lift a cup's worth of water above the water's surface, and as you lifted, it would get harder to pull it up. This is analogous to the vacuum—here caused by an incompressible liquid instead of nothingness—growing in strength as you do work to exert negative, sucking pressure on the cupful of water. If you let go of the cup, the "vacuum" that is the water in the tub will pull the cup back under, but if you pull it hard enough, it will destroy the vacuum, the water will return to the tub, and your cup will empty very, very quickly.

The gravitational effects of pressure are similarly unintuitive. Remember earlier when I told you that general relativity relates the curvature of space-time to the mass-energy density? Well, I didn't tell you everything. The truth is that GR relates the curvature of space-time to what is called the stress-energy tensor, which contains both the mass-energy density *and* the pressure.

Pressure creates gravity! Positive pressure creates attractive gravity, as with dark matter, and negative pressure creates repulsive gravity, as with dark energy. It's too much to show here, but it's as clear as day in the math.

Dark energy's slight negative pressure is imperceptible on the small scales of the Middle Realm, but it becomes dominant on cosmological scales where there is a lot of space-time. How large? On scales of one hundred million light-years and above!

At scales like these, we are beginning to consider our universe as a whole, and when physicists do this, we divide our universe's contents into three categories: radiation, matter, and space-time.

As we now know, radiation moves at or near the speed of light. In the parlance of physicists, this is "relativistic" stuff since it moves very quickly through space and almost doesn't move through time at all.

Light is a clear form of radiation, but, of course, not all light is equal. The energy density of the CMB, for example, which is ancient, having originated during the Big Bang, far exceeds the energy density of light from all other sources.

Another type of radiation is neutrinos, which are particles that have so little mass we can't measure it, even though we know it exists. As with light, most of which exists in the CMB, most neutrinos trace their origins to our very early universe. There are contributions from stars, galaxies, and radioactive decay, but the number of OG neutrinos far exceeds the portion created by these newer structures.

Another primary type of radiation is gravitational waves. The dominant source of gravitational wave energy is the orbital motions and collisions of black holes. Gravitational waves from our universe's birth have been predicted to have left an imprint on the CMB, like a kind of fingerprint, but these "primordial" gravitational waves are too feeble and swamped by noise to be measurable by our current technologies.

All of this radiation, added up, has a minimal energy density today. This wasn't always the case. For a short while, it was dominant. But nowadays, the energy density of all this radiation is no longer dominant. Currently, it's a thousand times smaller than the density of our second major category: matter.

Unlike radiation, matter moves slowly through space but rapidly through time. As we now know, most of the universe's matter is dark matter (80 percent), and the rest of it is ordinary baryonic matter (20 percent).

Both radiation and matter possess positive energy density and positive pressure. Both create attractive gravity. Space's expansion dilutes both as the universe's volume grows. Light is stretched. This results in radiation's energy density diluting more rapidly than that of matter. Since radiation is a wave tied to space, light is stretched to longer wavelengths and lower energies by the expansion of space-time.

At our universe's beginning, radiation's energy density was greater than that of matter. But after fifty thousand years, matter replaced radiation as our universe's dominant energy density. Of course, matter's dominance, too, was destined to end. After nine billion years or so, the third category of "universe-stuff" became dominant.

Space-time.

This substance, whatever it's made of, does not move through space *or* time, and its energy does not dilute. The energy of space-time is dark energy. Unlike the energies of radiation and matter, dark energy does not dilute with our universe's expansion, because it is the energy that is inherent to space itself. Make more space, and you make more dark energy. As our universe expands, the amount of dark energy also grows. (It's worth noting that recent results from the Dark Energy Spectroscopic Instrument have uncovered evidence that the density of dark energy is slowly falling even as its total amount increases with the expansion of space-time.)

Let's slow down for a moment and marinate on that last paragraph. It goes against everything you've probably learned about how energy works in our universe. There is no such thing as getting it for free. Perpetual motion machines are impossible.

Well, that's true in the Middle Realm. But it is *not* true in the Cosmological and Dark Realms. The rules here are different. Local energy is conserved, but global energy is not. Our universe spontaneously produces new space and, with it, new energy.

Although dark energy was dwarfed by the energy densities of

radiation and then matter in our early universe, it has now become our universe's dominant energy density. Our universe might have begun with the phrase "Let there be light," but it has since evolved to "Let there be dark."

■ ■ ■

While dark matter and dark energy provide compelling frameworks to explain these cosmological mysteries, they are not the only paths to understanding our universe's behavior at large scales. It could be that our current understanding of space-time, rooted in Einstein's general relativity, is incomplete. Perhaps the laws of gravity shift subtly—or dramatically—at cosmological scales, offering an alternative explanation for the deviant cosmic motions we see and that tripped us down this confounding path of discovery.

One prominent idea suggests that gravity might behave differently at extremely low accelerations or vast distances. In these scenarios, adjustments to Einstein's equations could mimic the effects of unseen mass or energy, eliminating the need to invoke dark matter or dark energy as separate entities. Instead, what we perceive as "dark" could emerge naturally from a more accurate theory of space-time and its interactions with matter and radiation, without requiring new quantum fields and particles (we'll get to these soon!).

A newly emerging argument in favor of so-called alternative gravity is that it better replicates the rate at which our universe's first galaxies formed, as recently revealed by the James Webb Space Telescope.

In the standard paradigm, galaxies form within massive halos of cold, dark matter that act as invisible scaffolding. These halos attract baryonic matter, causing gas to collapse, cool, and form stars. This process is gradual and hierarchical: Smaller structures merge over time to create larger galaxies. Grand-design spiral galaxies—the striking systems with well-defined arms—are believed to emerge relatively late in this model, typically when our universe was more than a billion years old.

When the Hubble Space Telescope (HST)—the most advanced telescope before the JWST—imaged the earliest galaxies it could reach, the galaxies always appeared clumpy and formless, similar to dwarf irregular galaxies and consistent with dark matter models. This is because the *visible* light the HST detects from the most distant galaxies is actually *ultraviolet* light that has been cosmologically redshifted into the visible range. Since almost all stars emit visible light, but only the hottest emit ultraviolet light, spiral galaxies, which appear beautifully detailed in visible light, are clumpy and discontinuous in the ultraviolet.

Once JWST, which detects a broad range of infrared light, began observing even deeper into the cosmos, astronomers were surprised to discover that grand-design spiral galaxies existed as early as one hundred million years after the Big Bang!

Alternative gravity models can cause galaxies to form earlier in our universe's history by enhancing gravity in low-acceleration regimes. As a result, baryonic matter collapses more rapidly under the influence of the modified gravity and stabilizes into disklike structures much earlier than in dark matter halo models.

However, these alternative gravity theories face a significant challenge: They suggest that general relativity might be a limiting case of a more fundamental theory incorporating additional principles or variables. This would mirror the relationship between Newtonian mechanics and relativity, where the former is accurate within a particular domain but fails under high velocities or strong gravitational fields. Such a shift would not necessarily discard general relativity but reframe it as a subset of a broader framework that governs the universe's structure and dynamics.

Remarkably, general relativity has succeeded in describing gravity across a wide range of scales and extreme conditions. From the motions of planets in our solar system to the behavior of light near black holes, general relativity not only explains the data but does so with a level of accuracy that inspires confidence. In cosmology, where subtle effects like gravitational lensing or the CMB's structure come into play, Einstein's theory is predictive. It holds up really, really well.

Any proposed modifications to general relativity must also explain the anomalies attributed to dark matter and dark energy and seamlessly replicate the immense success of general relativity in describing the universe. The more precise our observations become, the harder it is to accommodate the deviations that these alternative gravity theories require.

Even so, the door to alternative gravity remains open. The sheer scope of the challenges presented by dark matter and dark energy motivates us to continue exploring all possibilities.

Could these anomalies point not to exotic entities but to the need for a deeper, more profound, more comprehensive theory of space-time? This question drives the ongoing tension between established physics and revolutionary ideas, ensuring that the pursuit of answers remains one of the most vibrant and dynamic aspects of modern science.

For now, most physicists agree that thinking of dark matter and dark energy as "stuff" in space-time rather than a misunderstanding of gravity is probably more correct. But if these massive, cosmos-building protagonists *are* stuff, what are they composed of?

To have any hope of understanding that, we must turn away from galaxies, the CMB, and supermassive black holes. To dig deeper, we must shift the scale radically.

This means turning our gaze inward. Far inward to the very ingredients of existence itself: protons, photons, electrons, neutrinos.

Pack up your cat.

We're headed to the Quantum Realm.

QUANTUM REALM

Then Bohr retorted, so sharp and too true,
"Albert, please stop telling God what to do!"
The quantum world, while seeming askew,
Unravels her mysteries for the curious few.

—Andrew Corley

The cat is in the box, right? We may not know whether the cat is alive or dead, but we put that dang cat in there, so we definitely know that the cat is in the box.

Right?

If you're familiar with this thought experiment, you might already be nodding along. If you're not, don't worry; we'll pin down this boxed cat soon enough.

We've been grappling with the riddle of this enigmatic feline since the Austrian physicist Erwin Schrödinger first proposed it in 1935. That year, a figurative cat was placed into a box, along with a hammer, a vial of poison, and an atomic trigger. The box was shut, and that's when things started getting trippy. If the atomic trigger tripped due to just a single atom of radioactive decay, the hammer swung and broke the vial, and the cat died. But the box is closed. From the outside, we cannot know if the trigger released the poison. So is the cat alive or dead? Schrödinger argued that so long as the box remained closed, the cat inside was both dead and alive. At all times.

But there was a twist: While everyone debated the cat's state of mortality, not many asked whether the cat was in the box at all. So, was it?

Yes, it was. And no, it wasn't.

You see, Schrödinger's cat is a quantum feline. Even if the cat *ain't* in the box, the cat is *still* in the box. Perhaps, there is always a measure of "catness" everywhere at every location in the universe. Cats may be popping out here and there at all times. Heck, our cat may spontaneously transform into three mice, flutter into a pair of butterflies, or condense into the improbable seed of an apple.

This is just the slightest glimpse of the Quantum Realm's weirdness. At least weird to us Middle Realm dwellers. Totally ho-hum in the Quantum Realm.

The Quantum Realm is unlike anything we experience in the Middle Realm. Remember when your elementary school science teacher told you that matter could be neither created nor destroyed? That's Middle Realm talk. In the Quantum Realm, matter *can* be created *and* destroyed. Things can exist in two places at the same time. Physically separated particles can behave like a single object, no matter how far apart they happen to be. Particles of matter pop into and out of existence. And they shift from place to place without traversing the physical space between them. Here, space itself can lose meaning. In the Quantum Realm, observing an object doesn't just tell you something about it; it creates the reality you observe.

As weird as it may be, the Quantum Realm is the realm that manifests all others. If reality were an onion, then the Quantum Realm would be the core—the deepest, most essential layer upon which all others rely. We live our daily lives in the Middle Realm, a scale where objects are tangible and the laws of classical physics seem ironclad. But the Quantum Realm hums beneath, imperceptible and strange, weaving the fabric of the universe.

To put this into perspective, think of the Middle Realm as the world of objects we can see and touch, from a small chunk of matter to a pencil to a mountain to a star to a galactic arm. The Quantum Realm, on the other hand, is a world of dizzying smallness that begins at the size of an atom and descends into even more minute dimensions.

Imagine dividing a millimeter, that tiny tick on your ruler, into ten million equal parts. Each of those fragments would be an angstrom (10^{-10} meters), the scale of an atom. Now go smaller. Take one of those angstroms and split it a hundred thousand times further; this is a femtometer (10^{-15} meters), the scale of an atomic nucleus.

In more relatable terms, if a millimeter were the width of Earth, then an angstrom would be like a hot-air balloon, and a femtometer would be the size of a bacterium floating inside that balloon. Both angstroms and femtometers are incredibly tiny. The quantum particles that live here—quarks, electrons, neutrinos—are smaller still. But they're not nearly as small as our galaxy and the universe are big—relative to us, that is.

Go back to the first chapter of this book. The Middle Realm, where humans operate, is nestled between two extremes of scale—the incredibly tiny and the unimaginably vast. We have managed to stretch our minds and imaginations to picture the grandeur of galaxies and the observable universe, spanning 10^{20} to 10^{26} meters—scales that make human beings look like mere specks of nothingness. Yet it is somehow easier for us to wrap our heads around this immensity than to fathom the bizarre smallness and behavior of the Quantum Realm.

Why? You might think that zooming into the small should be as simple as zooming out to the large. But the quantum world isn't just a smaller version of what we know; it's an entirely different reality. On the largest scales, space and time behave in strange but still relatable ways. But in the Quantum Realm, the "rules" break down, and what we think of as solid and sure blurs into a haze of probabilities. The Quantum Realm isn't just small; it's a realm where even our fundamental understanding of "being" gets tossed into question. And that is why visualizing it is so much more challenging than picturing the giant sprawl of galaxies.

I have mentioned quantum particles throughout our journey. The most prevalent ones we've discussed are electrons, protons, and photons. As we've learned, quantum entities like electrons and photons sometimes act like tiny, localized objects, and sometimes they act like spread-out waves. All quantum particles share this characteristic.

More than a few readers have probably heard of the Heisenberg uncertainty principle; this is not a reference to the AMC series *Breaking Bad*! This Nobel Prize–winning idea by Werner Heisenberg is the quantum manifestation of a general trade-off we make whenever measuring the properties of waves, any kind of wave, including quantum waves. It holds that you can precisely measure the speed of a wave or the location of the wave, but never both simultaneously to a high degree of certainty. As you zero in on the value of one, the value of the other becomes fuzzy. This is because to determine a wave's speed, you must look over a range of distances, which makes you lose information about exactly where the wave is located. On the other hand, if you look at where the wave is located, you ignore this range of distances and therefore lose information about its speed. This is true for all quantum particles, which are fundamentally tiny waves.

Let's describe these tiny "particles" in a little more detail. I'm putting "particles" in quotation marks because, like light, which is massless, it's not very accurate to call these things that contain mass, well, things. They are the fundamental constituents of matter and energy that make up things, but they themselves are not exactly things. They are more like possible things, or, even more strangely, just mere possibilities.

There are six cardinal particles of the Quantum Realm that explain the world around us. There are two massless force-carrier particles: photons and their "sticky" cousins, gluons; these are classified as bosons. There are two particles of very small mass: electrons and neutrinos; these are leptons. And then there are the two high-mass particles called the up quark and the down quark; these are hadrons. Hadrons and leptons interact with themselves and each other by exchanging bosons. (For completeness, quark transformations are mediated by a set of three intermediate vector bosons.)

Quarks exchange gluons, thus forming bonds between them. When up and down quarks combine in sets of three, they form subatomic particles like protons and neutrons. As we've learned, massive protons interact with near-massless electrons by exchanging massless photons, and thus bond to form electrically neutral atoms, the most basic (and common) of which is hydrogen. Atoms then interact electromagnetically through the exchange of photons to form molecules and, ultimately, the macrostructures that make up the Middle Realm.

One thing to note is that all these quantum particles come in two types . . . twice! They are divided into particles and antiparticles, and they are also divided into real particles and virtual particles. First, let's talk about these real and virtual quantum things. Don't get tripped up by these terms: Both exist. The difference is that real particles last a long time. In contrast, virtual particles flit around and exist for only the tiniest fraction of a second. For now, understand that the exchange of photons that gives rise to the electromagnetic force involves *virtual* photons, not *real* photons. Likewise, there are virtual gluons in atomic nuclei that hold protons and neutrons together.

Since quantum particles are not truly particles that can be described by properties such as position and momentum, physicists describe them using a mathematical concept called the wave function. As the term suggests, the wave function embodies both discrete, particle-like behavior and more spread-out, wave-like behavior.

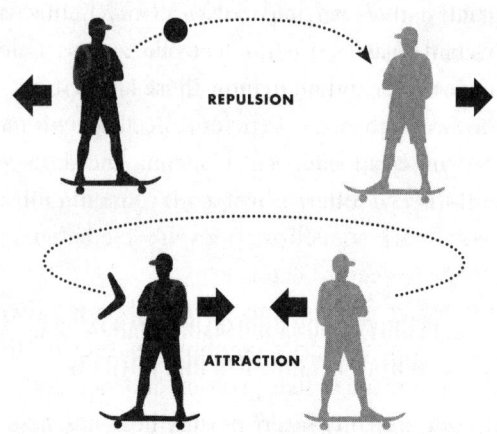

Fig. 17: Forces from Virtual Particle Exchange
All fundamental forces (except gravity) are mediated by virtual particle exchange; here attraction and repulsion are illustrated for the electromagnetic force.

Photons make up light and are the quantum particles we have the most direct experience with, even if we don't often appreciate the finer points of what it means to be a photon. The other quantum particles are a little more arcane. Imagine a tiny, invisible thingamajig that you cannot discern with any of your five major senses. You want to know what it is, so you do experiments to see how it behaves. You notice that sometimes it acts like a little particle. It bounces off other particles, sharing momentum, like colliding billiard balls or marbles. But other times, it acts more like a wave. When waves collide, they don't bump off each other. Instead, they overlap and interfere with each other in multiple locations simultaneously, combining to form a bigger or smaller wave. Under the right conditions, waves can even cancel each other out and annihilate one another.

An electron flowing through a semiconductor behaves like both a particle and a wave in just this way. In this case, the electric charges of individual electrons can be sensed by transistors.

However, when an electron flows through the semiconductor material, it can reflect off the "walls" of atoms, much like a wave, meet itself flowing in the opposite direction, and then combine to form standing waves that manifest as semiconductor energy bands. If electrons didn't do this, digital cameras of the sort we use every day would be impossible.

But here's the mind-boggling part: The electron doesn't just act like a particle and a wave at the same time; it can switch back and forth! It is unlike anything in the Middle Realm.

To see our electrons' strange behavior, we can repeat the famous experiment where individual electrons are shot at a barrier with two openings. When we don't participate in the experiment and actively observe which opening the electrons pass through, they seem to pass through both openings simultaneously, creating a wavy pattern on the other side. But if we insert ourselves into the experiment and actively monitor which opening the electron passes through, it suddenly acts like a particle, and the wavy pattern disappears. We can never directly observe the wave;

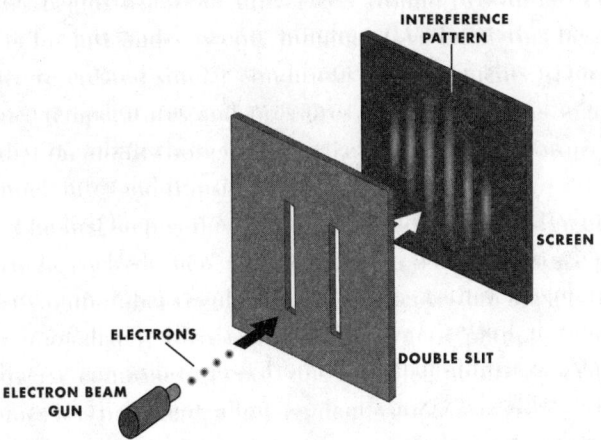

FIG. 18: DOUBLE-SLIT EXPERIMENT
Quantum interference pattern created by single particles, revealing wave-particle duality and the collapse of probabilistic states upon measurement.

we see only the result of wave interference. It's like the saying "Where there's smoke, there's fire." Here, the saying becomes "Where there's interference, there are waves."

Wave-particle duality, annihilation, superposition, tunneling, entanglement, and pair production are the phenomena that distinguish the Quantum Realm from the realms above it. Undergirding all of these phenomena are quantum fields, which, along with our OG universal protagonist, energy, are the true foundations of all reality.

Let's pause for a moment here. Words like "field," "superposition," and "entanglement" might feel like jargon from a language only physicists are fluent in. But we'll untangle each one, step by step. And once we do, I hope to provide you with the tools of quantum logic that will enable you to see and anticipate how the universe operates at this microscopic scale. But heads up: Though we physicists have wrangled with understanding the Quantum Realm for more than a century, few people (physicists included!) have achieved understanding. It's a funky-ass place, so don't get dejected if you feel torn or superimposed. Or entangled. Or annihilated. Just hang in there (like the cat in the box!).

Let's start with "superposition." Schrödinger put the cat, the poison, the hammer, and the trigger in the box and then never opened it. Then he wrote out equations explaining the cat's condition, and the equations state that the cat is in a mixture (a superposition) of being dead *and* alive—until you open the box and peek inside, that is. Then it is only either dead *or* alive.

Schrödinger's cat is meant to illustrate paradoxes inherent to the Quantum Realm. While unobserved, the cat, which is a stand-in for a quantum particle, is both dead and alive. It is in the box and outside the box. It is a cat and a not-cat. If it were a Cheshire cat and Lewis Carroll had been a brilliant Austrian physicist, it would be mad and not mad, smiling and not smiling, helpful to Alice and totally indifferent to her; until Alice looked at it, that is.

As with the cat, when quantum particles are unobserved, they are superimposed in multiple locations and/or states. But once observed—even in the most unobtrusive way possible, and not necessarily by a conscious observer, since, as far as we know, almost the entire universe completely lacks anything close to consciousness—the particle "picks" one state over another. It's either here or there. It's spinning either one direction or the opposite. It's either going somewhere or it isn't. Until then, it's just a little packet, a little quantum, of possibility. We can never directly observe the multiple-states state, but we know it is real.

Let's go from a quantum cat to a quantum car. Suppose our quantum car can move at only five different speeds—ten, thirty, fifty, seventy, and ninety miles per hour. When unobserved, our quantum car moves at a mixture of all five speeds simultaneously. But once observed, the car's speed is determined, and it will be found to be traveling at only one speed. When multiple measurements are performed, every predicted speed is observed at precisely the frequencies predicted; in our hypothetical, the car will be moving at fifty miles per hour 50 percent of the time, at either thirty or seventy miles per hour 15 percent of the time, and at either ten or ninety miles per hour 10 percent of the time. This is superposition in action.

Most theories in physics involve the mathematics of change, also known as differential calculus. In the realms above the Quantum Realm, the physical quantities that differential calculus deals with make intuitive sense to us: speed, momentum, position, energy, and so on. But in the Quantum Realm, the thing that changes is something for which we have no intuition. This is the "wave function," which was discovered (or imagined, or created, or deduced, or conjured) by the same Erwin Schrödinger. In the simplest cases, the wave function comprises a simple sum of quantities, each representing an allowed state of the particle, multiplied by a number related to the probability of that state being observed. (This is exactly the same as our quantum cat and its deadness-aliveness and our quantum car and its five speeds.)

Here's another example: a coin flipping in the air. While the coin is airborne and spinning, we can say it is simultaneously both heads and tails. We can describe this superposition state probabilistically by saying that it has a 50 percent probability of landing heads or tails. It is not until we catch the coin and look at it that it becomes definitively one or the other.

One of the strangest consequences of superposition is that one of the multiple states that a particle can be in is two different locations. So, a true quantum description of our cat may describe it as being both alive and dead *and* both inside and outside the box. The phenomenon of quantum tunneling allows us to calculate the exact probability of finding the cat in locations outside the box.

In our everyday experience, objects cannot be in two places at one time. Cats, or cars, or people don't walk through walls. So why do physicists accept that our quantum equations describe actual reality? The answer is that these ideas lead to accurate predictions of what we observe in our experiments; it's no different from when Copernicus realized Earth orbited the Sun and not the other way around, even though he had no means to zip out to space and confirm it by observing Earth and the Sun from, say, the orbit of Jupiter.

In fact, the most accurate and precise predictions in all of science come from the Quantum Realm. The mathematical discipline of the Quantum Realm, known as quantum mechanics, provides predictions with extraordinary accuracy and precision, often exceeding the experimental capabilities of measuring them. Where we *can* measure them, in experiments involving atomic and molecular spectra, the predictions of quantum mechanics can be accurate to within a few parts in a trillion or better.

One is tempted to conclude that describing the Quantum

Realm in this almost magical way is just a mathematical trick. But while superposition may initially appear counterintuitive, its experimental validation and successful applications in technology (such as quantum computing and quantum cryptography, which exploit the ability of quantum systems to exist in multiple states simultaneously to perform computational tasks or secure communications) demonstrate that it is not merely a trick. It is a fundamental aspect of the Quantum Realm.

A strange phenomenon closely related to quantum superposition is quantum entanglement. Here, instead of one particle being in multiple states, multiple particles can behave as if they were one. This paradoxically leads to the appearance of either space being rendered moot or faster-than-light communication.

Let's return to our flipping coin. Suppose we have two entangled coins: When one is observed to show tails, the other must show heads, and vice versa. Each coin is individually in a state of superposition. While they're flipping, they're simultaneously both heads and tails. We cannot know what either coin will be before we catch one and look at it. But we know two things. First, each coin has a 50 percent probability of being heads or tails once observed. The second is that when we stop one from flipping and look at it, the probability of the other being in the complementary state is 100 percent. They're correlated. And that correlation remains, regardless of how far apart they happen to be. For these coins, it's as if space does not exist. Crazy. Unexpected. But quantum entanglement, like quantum superposition, has been experimentally tested, verified, and incorporated into technologies we use every day.

■ ■ ■

Quantum particles are like tiny characters in a grand cosmic play, each with a unique set of traits that determine how they interact with others. These traits, or "properties," define the rules of their behavior and their role in our universe. These traits include their rest mass, spin (a kind of intrinsic angular momentum),* and various types of charges, which act like the rules of attraction and repulsion between particles.

This brings us to particles and antiparticles. The difference between the two types lies in having opposite particle properties: either opposite charge or opposite number.

We're already familiar with electric charge, which comes in two types: positive and negative. It's like a magnet, where opposite poles attract and like poles repel. But where it concerns quarks and gluons—the building blocks of protons and neutrons—there's another kind of charge called color charge. Both quarks and gluons possess color charge.

Despite the name, color charge has nothing to do with actual colors. Instead, it's a metaphor used to describe how these particles interact through the strong force, one of nature's four fundamental forces. If electric charge is like a two-way street, with positive going one way and negative going the other, color charge is more like a three-lane highway. It comes in three varieties labeled "red," "green," and "blue." These labels are symbolic, chosen to help us understand how quarks combine and balance their charges, much like mixing primary colors to make white light.

For example, three quarks—one "red," one "green," and one "blue"—combine to form a proton or neutron, in a perfectly "color neutral" way. This balancing act is governed by virtual gluons, which are the messengers of the strong force. They act like elastic bands holding quarks together, constantly exchanging their own color charges to ensure the particle stays stable. In

---

* Spin logically translates to rotation, but that's not what this is. Angular momentum in the Quantum Realm is not about a particle spinning like a top. Instead, it's an intrinsic property that particles possess, even elementary particles that have no physical size. Every electron, for instance, has exactly the same amount of spin (spin-½), which can be oriented in different directions.

this way, color charge is the glue that binds the atomic nucleus, giving matter its structure.

Antimatter particles (also called antiparticles) have opposite electric and color charges from their matter particle counterparts. Whereas an electron has an electric charge of −1, an antielectron (also known as a positron) will have an electric charge of +1. Since quarks have both electric charge and color charge, antiquarks have the opposite values of electric and color charge from their quark counterparts. Particles also have more esoteric properties known as lepton number or baryon number. Neutrinos possess no electric charge, but do have a lepton number, allowing them to have an antiparticle. The lepton number equals +1 for regular neutrinos and −1 for antineutrinos. Likewise, the antiparticles of electrons and quarks, respectively, possess the opposite lepton and baryon numbers.

There are two essential things to know about antimatter. First, we ain't got none! Our universe is composed entirely of normal matter with almost no antimatter as far as we can see. Second, when matter and antimatter come into contact, they annihilate completely. Their energy ceases to be constrained in the matter quantum fields (we'll get to fields soon!). It's sent into the electromagnetic quantum field and converted to light. The particle's mass has been destroyed, and this is how we know there are no significant concentrations of antimatter in the universe. It would not be possible for this antimatter to avoid matter forever. This means that, from time to time, we should see big bursts of high-energy light coming from locations where matter and antimatter are meeting and annihilating. But we don't. It ain't there.

Now consider the incredible coincidence that allows stable atoms to form. Electrons possess an electric charge that is precisely equal and opposite to that of the proton. I find this astonishing. A proton is a composite particle made up of even smaller up quarks and down quarks, which are held together by a swarm of massless gluons. An electron, by contrast, is just an electron. As far as we can determine, it is fundamental and cannot be further subdivided. Also, recall that protons and neutrons are almost two thousand times more massive than electrons.

To illustrate how preposterous the canceling of a proton's charge by an electron is, consider an analogy with sound. Noise-canceling headphones work because sound waves can collide and cancel each other out. A small microphone in the earphone picks up the ambient noise and then plays the "opposite sound," silencing the ambient noise in real time.

Now imagine a tiny creator of sound—an electron—precisely canceling an ensemble—a proton—of much larger sound creators. For the sound corresponding to our electron, we have a flea's fart. For our proton, we have a sound created by a musical trio of two trumpets and a tuba, accompanied by thousands of snare drums (gluons). Imagine that the flea's flatulence matches the ensemble's volume but with the opposite sound, resulting in silence even as the instruments are played at maximum volume. This is precisely what the electron's charge does in canceling out the proton's. Crazy!

Another strange characteristic of each type of quantum particle is that we can't distinguish one from the other. Every electron is exactly the same as every other electron; every up quark is exactly the same as every other up quark. It's as if they've been cloned. To contrast the strangeness of quantum particles' identical nature with something familiar, consider cells in your body. Unlike quantum particles, cells are not all identical. A skin cell differs from a neuron, and even within a single type each cell exhibits slight differences due to its unique position, environment, or minor mutations. These variations allow for specialization and diversity, which are essential for life as we know it.

In stark contrast, quantum particles of the same type are indistinguishable in every measurable way. You can't point to any unique feature that sets one apart from another. Particle indistinguishability is *the* defining feature that gives rise to quantum statistics—it's not a minor quirk, it's the engine under the hood. In classical physics, you can slap a name tag on every particle and track them individually, like kids on a field trip. But in the Quantum Realm, when dealing with identical particles (like electrons or photons), that labeling scheme literally and figuratively collapses. Treating indistinguishable particles as distinguishable is like trying to play chess with poker rules: It breaks the game. To give just one example, phenomena like electron flow in computer chips—responsible for all modern electronics—wouldn't be possible without quantum indistinguishability. So, while cells thrive on differences, quantum particles depend on uniformity to enable the universe's fundamental processes.

This indistinguishability of quantum particles also hints at a deeper, more profound truth about reality: The particles are not truly the fundamental entities. As far as we know, these quantum particles are not reducible to even smaller things. But that does not mean that these particles lie at the absolute rock-bottom foundation of all reality.

Consider music again. Quantum particles are like musical notes. Musical notes exist. And they're identical. A C-sharp is a C-sharp. A B-flat is a B-flat. But the sound we hear is a consequence of our eardrums sensing air vibrations. Air is the medium vibrating, and the instruments or voices that make it vibrate are the actual reality. In the case of quantum particles, *quantum fields* are the medium—the air of the Quantum Realm—and energy is the musician, the initiator of the particles' existence. What we call quantum particles are better described as excitations of quantum fields. The definition of "excitation" here is "energy has been added." This is virtually equivalent to musical notes.

You are a symphony of quantum fields!

It's not an exaggeration to say that reality is nothing more than energy causing vibrations across an invisible mesh of nebulous quantum fields that exist everywhere at every moment at every location in the universe. More accurately, these fields and the energy that courses through them *create* everywhere.

Quantum fields are the universe's hidden foundation. They determine how particles behave, interact, and even come into existence. These fields are the reason the world on the tiniest scales can seem strange and counterintuitive compared with our everyday experiences.

Imagine space—outer space, the space around you, the space within you—as not just empty but thoroughly permeated by invisible "fields." These fields aren't like the fields you see in a meadow; they're more like a hidden, dynamic fabric woven with countless interlaced threads. Picture these threads as vibrating strings on an infinite harp, each representing a quantum field. Every type of particle—electrons, quarks, photons—has its own corresponding string, tuned to its unique frequency. (Please don't confuse this analogy with string theory!)

Here's the crucial idea: Particles are not separate from their fields. Instead, they are "excitations" or "ripples" in these fields, like notes played on an instrument's strings. A particle is like a vibration in its corresponding field—a musical note, if you will. When it's in motion or interacting with others, the symphony grows, with ripples and harmonies traveling throughout the fabric of reality.

Those virtual particles I mentioned earlier? Those are due to tiny fluctuations in the quantum fields of each particle type. Imagine looking at the string on one of our instruments. It is calm and not vibrating, producing no obvious waves, because it has not been plucked. But if you were to zoom in and look at this string with a powerful microscope, you'd see that it is anything but calm. There are tiny fluctuations that exist along the string on very short timescales. The string is never completely at rest. Our quantum fields are similar, and their tiny fluctuations give rise to virtual particles.

Quantum fields also give rise to quantum superposition. And

quantum entanglement. And annihilation, tunneling, and quantization. In a sense, there are fields within fields. Yes, there are quantum fields for each particle, including virtual particles, but there are also fields for each *property* of quantum particles, and the interaction of these fields is essentially what defines the particle in question.

Suppose there was a field responsible for cans of soda. When enough energy gets added to this field—poof!—you get a can of soda. But energy coursing through this field also contains information *about* the soda can. Sometimes, the can is a tall boy. Other times, it's a pony can. Sometimes, the can contains Coke; other times, it contains Mr. Pibb. Sometimes, it even contains plain old seltzer (yuck!). Sometimes the can is really cold; other times it's room temp (more yuck!). The wild thing to understand is that the Soda Can Field makes cans of soda, but it also interacts with other fields that make judgment calls about the can, which we then use to describe the can and its contents. The Soda Can Field has corresponding fields that pertain to the size of the can, its temperature, what kind of soda is inside, whether you got that soda at Whole Foods or the bodega, and so on—and all of these are independent of each other and of the can itself, but they also all interact.

The interaction between quantum particles in the various quantum fields happens when they exchange quanta of their *respective* fields. For example, two real electrons, which are excitations in the quantum electron field, repel each other by exchanging virtual photons, which exist in the quantum electromagnetic field, the field responsible for interactions between real electrons and protons. This exchange of virtual photons is what we perceive as the electromagnetic force that causes the repulsion between the two negatively charged electrons.

The quantum fields that define the electron and its charge are like the microscopic gears of a finely tuned machine, while the classical electric field is the smooth outward motion we see as the result. Quantum electromagnetic fields are the origin, and the classical electric field is the outcome. The classical electric field can be viewed as an averaged effect of the underlying quantum electromagnetic field when observed from a distance and in a calm, low-energy environment. At this scale, the quantum ripples smooth out, leaving us with the familiar electric field that obeys the well-known equations of classical physics.

Imagine placing a lone electron in a space free from other real particles. Its charge modifies the local quantum electromagnetic field in its vicinity, creating a force field that can influence other charged particles that enter its vicinity. This field tells us how the electron "reaches out" to affect its surroundings. If you were to place a tiny test charge nearby, it would "feel" a push or pull depending on its charge and position. We observe and measure this classic electric field in experiments.

The strong nuclear force, on the other hand, is mediated by virtual gluons, which are excitations in the quantum gluon fields. (There are multiple types of gluons, so there are also multiple types of gluon fields.) To form protons or neutrons, gluons are exchanged continuously between quarks, creating a "gluon bridge" between them. These gluons are virtual particles, meaning they appear not as actual particles we can detect but as particles that temporarily mediate the interaction. Imagine two quarks, Quark A and Quark B, that are relatively close to each other. As Quark A emits a virtual gluon toward Quark B, Quark B then absorbs this virtual gluon and responds by emitting another virtual gluon back toward Quark A. This back-and-forth exchange of gluons *is* the strong nuclear force.

As we know by now, the world we see around us—the Middle Realm—is governed by two main fundamental forces: gravity and electromagnetism. These forces weaken as objects move farther apart, which is why the Sun's gravity exerts less pull on distant planets and why a regular magnet can't attract an object from across the room.

But in the Quantum Realm, things are very different. Here, the strong nuclear interaction is the dominant force. Instead of weakening with distance, the strong nuclear force grows stronger as quarks try to move farther apart. It's as though quantum elastic bands connect them. This behavior is called confinement, and it is why quarks are never found alone; they are almost always grouped into pairs or trios. (Protons consist of two up quarks and one down quark, while neutrons are made of two down quarks and one up quark.)

Another consequence of confinement is that gluons remain inside their proton or neutron and do not participate in binding neutrons and protons to each other. For this, we utilize the concept of virtual mesons, which are composed of two quarks. When protons or neutrons are close, they exchange these virtual mesons, binding the particles tightly to form a nucleus.

As I've mentioned, electrons, unlike protons and neutrons, are considered indivisible particles, yet they exhibit some surprising behaviors in certain materials, where their properties seem to split apart. Think of an electron as carrying two key traits: electric charge and spin. In some conditions, these traits can be separated into the holon (carrying the charge) and the spinon (carrying the spin).

Beyond this, electrons can also exhibit a property related to their orbital motion around a nucleus. This motion, or "dance," can detach from the electron in certain quantum materials, creating a particle-like excitation known as an orbiton. These examples—holons, spinons, and orbitons—demonstrate how electrons, like other quantum particles, can be divided into distinct behaviors, highlighting the strange and unintuitive nature of quantum fields and the particles that exist within and around them.

These phenomena are the basis for all ordinary matter from your shoes to the Sun to the Milky Way. And all of these phenomena—quarks confined by gluons, mesons gluing nucleons together, and electrons fractionating into holons, spinons, and orbitons—are possible because of quantum fields.

Quantum fields also provide us with our current leading candidate for what "makes" dark matter: the axion, a hypothetical and thus far unobserved quantum particle. If axions exist, they will be incredibly light, interact only weakly with other matter, and move slowly—three properties that make them an ideal dark matter candidate. Axions wouldn't emit, absorb, or scatter light, but their mass and collective gravitational pull could help shape the universe's large-scale structure.

Physicists first proposed axions not to solve dark matter, but to resolve a puzzle with atomic nuclei—specifically, why protons don't exhibit a certain asymmetry in their electric field that theory predicts they should. The introduction of the axion field would cancel out this asymmetry. Later, researchers realized that the same axion, produced in vast numbers in our early universe, could account for dark matter.

Experiments such as the Axion Dark Matter Experiment and the CERN Axion Solar Telescope are actively searching for evidence of axions converting into photons in the presence of magnetic fields. So far, the search has come up empty, but the theoretical allure remains strong. If axions are real, they may not only solve an old quantum riddle but also help explain what makes up most of the matter in the universe.

Whether or not axions are the answer, they show how quantum fields continue to offer elegant solutions to the universe's biggest mysteries. Indeed, they are also key to understanding mass itself. For example, the proton's mass is not primarily due to the intrinsic masses of its constituent quarks, which account for only 1 percent of a proton's mass. The other 99 percent is due primarily to the *massless* virtual gluons that these quarks exchange. Their energy, when confined within the proton, contributes to its overall mass, as described by Einstein's equation

$E = mc^2$. This confined energy endows the proton with its (very small) gravitational and inertial properties, illustrating how mass arises from the fundamental interactions of quantum fields.

Ultimately, quantum fields and the energy that excites them appear to be the most fundamental components of reality. Everything else—from particles to forces, atoms to molecules, stars to GMCs—appears to emerge from them. However, there is an ongoing debate in physics about whether space and time are also fundamental or if they, too, emerge from something deeper.

My thinking, which I have not seen anywhere else, is that the existence of fields necessitates the existence of space, while the existence of energy requires the existence of time. Fields require a geometric framework, which we experience as space, and energy's manifestations require a before and after, which we experience as time.

These ideas are reflected in the Heisenberg uncertainty principle, which links uncertainties in position and momentum, or energy and time, to the Planck constant, $\hbar$. This universal constant governs the energies that light can possess, the lifetimes of virtual quantum particles, the smallest distances possible, the earliest time that can be measured, all motion, and much more. The fact that we have been able to discover this universal constant is miraculous. But that's what the rigorous application of the scientific method over centuries yields: miracles.

Most crucially, the Planck constant defines the smallest possible action in the universe. Through this principle and others, we see that space and time are deeply connected to fields and energy, suggesting that they are just as fundamental to the ultimate structure of reality.

This may sound pedestrian—of course, space and time are fundamental to our reality—but to be non-emergent means that space and time are fundamental to *all* reality. This idea—that space and time are intrinsic to all reality—opens the door to some of the Quantum Realm's most profound mysteries. To explore these, we must venture into the most extreme environments known: black holes.

▪ ▪ ▪

Black holes lie at the interface of the Quantum, Middle, and Cosmological Realms, bridging our understanding of fundamental physics and the universe's behavior across all scales. Black holes are derived from the mathematics of general relativity, which is couched exclusively in the language of energy and geometry. The equations predict two important boundaries of a black hole—its center and its edge. At both, quantum effects are thought to be dominant.

The center of a black hole is described as an infinitesimally small region where space-time curvature and energy density approach infinity. Here, the laws of physics as we currently understand them break. We call this point the singularity. Unfortunately, we cannot gain a proper understanding of the quantum effects at the singularity, because that would require a physical theory that some call the holy grail: a theory of quantum gravity. And we ain't got one.

At some finite radius away from the singularity is the black hole's edge—the event horizon. This is an intangible boundary at which nothing, not even light, can escape the black hole's gravitational pull. The event horizon marks a clear dividing line between a black hole's interior and the exterior universe. Inside, it's as if time and space have exchanged roles: Outside a black hole, we can only move one direction through time: forward. Inside a black hole, we can only move one direction through space: toward the singularity.

The radius of a black hole's event horizon scales with its mass. This result stems directly from the equations of general relativity and has been confirmed by observing how its gravity affects nearby matter and light. Doubling a black hole's mass doubles its radius; tripling the mass triples the radius. A black hole with the mass of the Sun would have a radius of only 1.8 miles, or roughly 3 kilometers. An Earth-sized black hole would have a mass one thousand times that of the Sun. The supermassive black hole at the center of our galaxy, equivalent to about four million solar masses, has an event horizon radius that is only about 120 times the Earth–Sun distance.

Unlike at the singularity, we do have theories that allow us to predict quantum effects at the event horizon—so-called black hole thermodynamics. According to the physicist Stephen Hawking, black holes are not entirely black: Just outside the event horizon, they emit radiation, which we call Hawking radiation. This emission imparts a temperature to the black hole, known as the Hawking temperature. This is extremely cold and is inversely proportional to the black hole's mass. A small black hole with one solar mass would have a Hawking temperature of around a billionth of a degree Kelvin. Supermassive black holes are many times colder than that, without ever reaching absolute zero.

This thermal radiation, as cold as it is, results from the quantum nature of fields near the event horizon and the interaction between quantum mechanics and gravity. Inside the black hole, the singularity converts energy to space-time curvature. But at the event horizon, the black hole converts space-time curvature to energy, which is ultimately lost as Hawking radiation. This radiation causes the black hole to gradually lose mass and, over an extremely long timescale, evaporate completely.

Currently, our universe is not nearly old or cold enough for black holes to have lost any mass by this mechanism. All black holes in the universe today ingest more energy than they radiate, meaning that they are all still gaining mass even as they also emit some Hawking radiation.

The most isolated black hole you can imagine in the deepest void of space is still embedded in a sea of background radia-

FIG. 19: BLACK HOLE STRUCTURE
Anatomy of a black hole: event horizon, singularity, ergosphere (for rotating black holes), and potential accretion disk.

tion of all sorts. Until the universe reaches a critical temperature threshold that is really, really cold, all black holes will continue to grow. Only once the surrounding universe is cooler than the black hole itself—say, less than a billionth of a billionth of a degree Kelvin—will it begin the process of evaporating, and ultimately disappearing, through Hawking radiation. Current estimates put this time frame at least a million trillion years into the future—and our universe is only 13.8 billion years old. We have a *long* way to go.

A peculiar theoretical property of black holes that connects them to the Quantum Realm is something every self-respecting sci-fi geek has a dreamlike inkling of: Black holes can be entangled. With these "wormholes," two event horizons can share a common black hole interior, as if two doors on opposite sides of the universe lead into the same room.

Another characteristic tying black holes to the Quantum Realm is their descriptions. Remember how I mentioned earlier that all quantum matter particles of the same type are indistinguishable from one another? This means each one can be described by a short list of properties: mass, spin, charge, and quantum numbers. The same holds for black holes. Each one can be described by its mass, spin (angular momentum), and electric charge. Black holes with identical properties are indistinguishable, just like quantum particles!

Black holes bridge our understanding of fundamental quantum physics at the smallest scales with the universe's behavior on the largest scales. They intersect with quantum mechanics, classical physics, and cosmology. They challenge our current theories and inspire the search for a more comprehensive framework that can explain the behavior of matter, energy, and space-time under the most extreme conditions in the universe. The ultimate expression of which was at the beginning of the universe itself, when time was but a babe.

Enter the Temporal Realm.

TEMPORAL REALM

Dr. Manhattan: There is no future. There is no past. Do you see? Time is simultaneous, an intricately structured jewel that humans insist on viewing one edge at a time when the whole design is visible in every facet.

—Alan Moore, *Watchmen*

When our universe began, you were reading this book.

I know. Sounds like another one of my absurdities. But stick with me. This is the story of how my relationship with time underwent a radical transformation across three defining moments in my life, ultimately leading me to the Temporal Realm. This is where time itself becomes the central character, as real and consequential as mass, space, and energy are to the other realms we've visited so far.

I was ten years old when my perspective on time first shifted. I'd been reading our family's *World Book Encyclopedia* and stumbled across the entry on Einstein and his idea that time is relative. Eight years later, it shifted again when I was in the U.S. Navy. After boot camp, I was assigned to a program run by the U.S. Marines. Marines don't play with time. Being "on time" meant showing up fifteen minutes early . . . or else. Civilian time was a suggestion. Marine Corps time was a commandment. It changed how I operated.

The third moment came eighteen years later, in 2003, during

a visit to the University of Chicago. There, I met the physicist Rocky Kolb. In one of our conversations, he mentioned the apocryphal Mark Twain quotation that begins the Middle Realm chapter: "It ain't what you don't know that gets you into trouble. It's what you know for sure that just ain't so."

That line messed with me.

I started questioning my entire understanding of time. I realized I'd been treating my perception of "now" as the one true moment. But what is "now" really? Is it a universal feature of reality, or just a trick of consciousness?

We treat the present moment as if it were the definitive, unfolding edge of the universe. We assume "now" is the place where reality happens. The past is gone, the future is on its way, and now is the only thing that's real.

But the more I thought about it, the more that story broke down.

After all, every human being who has ever lived or ever will live has experienced or will experience their version of the present moment. So, whose "now" is the right one? Is there even such a thing?

That's when I learned about *eternalism*, which asserts that all points in time are equally real. Past, present, and future exist simultaneously, much like different locations in space. Just because you aren't standing in Chicago doesn't mean Chicago ceases to exist. Likewise, just because you're not living in the year 2525 doesn't mean that year isn't real *right now*.

This is the heart of the Temporal Realm, the conceptual landscape where time is not just a flowing river but a dimension like height or width. It's a realm where simultaneity, causality, and change all get redefined. Unlike the Middle Realm, where time feels steady and predictable, or the Quantum Realm, where time almost disappears, the Temporal Realm invites us to rethink what it means for events to happen at all.

If eternalism is correct, then my initial statement is valid: You are reading this book at the moment of our universe's creation. You're reading it at the heat death, trillions of years away. And you're reading it now. Every moment exists in an eternal, infinite present. And maybe, in some sense, so do you.

We explored an unintuitive subtlety of time in the Cosmological Realm, where Einstein's famous thought experiment demonstrated that simultaneity is relative: Two events that occur at the same time for one observer—defining their now—can happen at different times for another, depending on their relative motion. But this isn't just a strange artifact of high-speed trains and lanterns; it scales up. When you stretch that idea to cosmic distances, it takes on a much more unsettling form. Consider the Andromeda paradox.

Formulated independently by C. Wim Rietdijk (1966) and Hilary Putnam (1967), and popularized by the Nobel Prize–winning physicist Roger Penrose, the Andromeda paradox is a thought experiment that forces us to confront how warped our intuition about "now" really is.

According to relativity, "now" isn't a universal feature of the cosmos; it's personal. Every observer carries their own unique set of events they consider to be happening simultaneously. This set of events forms what is called their plane of simultaneity, a cross section of our universe that each interprets as the present moment. And here's the kicker: That plane tilts depending on how you move.

Let's break it down.

Imagine two people, Kaden and Kaori, standing together on a city sidewalk. They both glance up at the night sky and agree that the Andromeda galaxy, 2.5 million light-years away, is heading toward us. "That's what's happening now," they say, pointing toward it.

But then Kaori starts walking toward the galaxy, while Kaden strolls in the opposite direction. Because of their motion, relativity tells us that their planes of simultaneity tilt ever so slightly in opposite directions. And that tiny tilt, multiplied across millions of light-years, makes a big difference. To see how this works, let's return to Einstein's train thought experiment but crank the dial to infinity.

Originally, Einstein asked us to imagine a passenger sitting at the center of a train car, holding a lantern. There's a door at each end of the car, and the light from the lantern opens each door when it reaches it. Because the passenger is equidistant from the two doors and in constant motion with them, they see the light hit both at the same time. Meanwhile, an observer standing outside the train on the platform sees the rear door of the train moving toward the light and the front door moving away from it. From their point of view, the light hits the rear door first, and then the front door.

Now imagine that train isn't just one car; it stretches infinitely in both directions. Car after car, no end in sight.

The passenger in the center of one car sees the doors at each end of their car open at the same moment. They also see the doors in the next car forward and the next car back open simultaneously. To them, this simultaneity continues forever. It's as if a wave of "now" were sweeping through the train, with every pair of doors, fore and aft, opening in perfect sync.

But someone on the ground perceives these same events separated by increasing delays with distance; they're not simultaneous at all. They see the rear doors, moving toward the light, open first. The front doors, moving away, open later. When the light reaches the next pair of cars, that time difference is compounded. If they saw a two-second delay between the doors on the first pair of cars, they would see a four-second delay on the second pair of doors, a six-second delay on the third pair, and so on. The desynchronization scales linearly with distance.

Go far enough down the train, and what one observer sees as two doors opening at the same moment, another sees as events separated by minutes, hours, even days. Stretch the train to truly cosmic scales—say, millions of light-years—and the difference becomes mind-boggling. The passenger might observe doors opening simultaneously throughout the entire train, while the outside observer could perceive these same events as separated by years or centuries.

The takeaway is profound: "Now" isn't just slightly different for observers in different reference frames; it can be radically different, with the gap growing proportionally to distance.

If different "nows" can coexist, if each observer's present is just one among infinitely many, then perhaps all those moments do exist all at once. That's the gateway to eternalism, the view that past, present, and future are equally real, and "now" is just a trick your brain plays as you surf a tilted slice of space-time.

A second school of thought regarding the nature of time is *presentism.* In this view, time is a sequence of unfolding events that exist only in the present moment. There is no past and no future, only the present. As soon as it happens, it's gone, and we're in the next moment, and then the next, and then the next, and so on, ad infinitum.

In 2024, Stephen Wolfram, a renowned British American physicist (renowned to nerds, at least), proposed a hypothesis of time that supports presentism. According to Wolfram, time consists of a series of discrete steps, in which the laws of physics iteratively evolve the universe as it goes from one configuration to the next. This approach resembles certain computational calculations that require step-by-step progressions. Such problems differ from those where a single equation describes a phenomenon (like the speed of Newton's falling apple over time) and provides an immediate solution. In the case of the apple, you can pick any moment in time and use a single equation to determine how fast the apple is falling. In step-by-step problems, on the other hand, you can't just plug in parameters to know the outcome. You must calculate each step before the next can emerge.

These steps do not coexist but manifest sequentially, aligning with the core idea of presentism.

While there is debate about whether eternalism or presentism is the correct philosophy, or whether time is objectively fundamental or an emergent phenomenon, the data show that our universe has a finite age. It had a beginning (for the most part—we'll get there). It may very well have an end. Currently, we're somewhere between these two.

The Temporal Realm of the universe is the dimension of time, which, along with space, forms the fabric of space-time. It allows events to unfold, apparently sequentially, creating the perception of a past, present, and future. In general relativity, time is not an absolute entity but is relative, shaped by the presence of mass, energy, and the motion of observers. This interplay means that time flows differently depending on gravitational strength or velocity, leading to phenomena such as time dilation.

Philosophically and theoretically, questions about the nature of time remain among the most profound we can ask. Is time infinite, or did it begin with our universe? Could it be cyclical, looping through endless repetitions, or will it stretch endlessly into the far future, never to return? These aren't just abstract musings; they cut to the heart of what time is, and whether it's fundamental or emergent. They also tie deeply into our understanding of space-time curvature, dark energy, and our universe's ultimate fate.

To truly explore the Temporal Realm, we must grapple with time's origin. If time is a dimension, how did it arise? Was there ever a moment when time didn't exist? That question immediately draws us toward our universe's origin, because time, as we know it, is inseparable from the expanding fabric of the cosmos.

For most of the twentieth century, physicists debated not only how our universe began but also when matter came into existence, and implicitly, when time itself began to tick. There were competing ideas about the respective origin of the universe and matter, just as there are currently differing philosophies about time. However, in 1990, the debate was settled, and the inflationary Big Bang model emerged as the clear winner.

Today's competing models of the universe's origin are primarily variations on the Big Bang, with its three nonnegotiable foundational ideas:

- Matter and energy came into existence in a single event.
- The newly created matter and energy were extremely hot, dense, and uniform.
- Space has been expanding at various rates ever since, driven by that initial impetus and modified by the changing energy densities of the universe's constituents.

With those three nonnegotiables in mind, it's clear why the term "Big Bang" stuck, even though no explosion occurred. What we're really describing, however, is less a moment of birth than one of immense transformation: a sudden shift in the nature of reality, in which energy, matter, space, and, yes, time all sprang into existence or underwent radical change. From this origin, all other instants in the universe cascade forward.

And what triggered this transformation? Energy. A surge of it, powerful enough to drive not just the creation of matter and light, but the expansion and curvature of space-time itself. This energy might well have come from the quantum fields that underlie all of physics.

Perhaps our universe's story is best understood as the ongoing evolution and exchange of energy between fundamental quantum fields. If this view is correct, and I contend that it is as correct as any other, then energy isn't just a feature of the cosmos; it is as essential as the quantum fields themselves. The continuous exchange, transformation, and redistribution of energy across fields and particles are what give the universe its existence, structure, motion, and time.

However, that brings us to an even deeper mystery: Where did this energy originate? Our universe's inception demands an energy source, yet no clear natural candidate presents itself. This isn't just a missing puzzle piece; it's a hole in the foundation. Without identifying the origin of energy, we have an incomplete understanding of our universe's birth and subsequent evolution. This question touches on everything because it explains why there is *something* rather than *nothing*. Indeed, why do we exist?

As physicists, we confront this chasm in the same way we always have: by combining observation, mathematics, and imagination to hypothesize new aspects of reality. Could the missing energy source be a new quantum field waiting to be discovered? Could it be the mechanism that triggered both time and energy into being?

In the following pages, I'll introduce a field that can do just that: the inflaton field. Along with its wild companion, cosmological inflation, it offers a compelling scenario in which our universe's time, space, and energy are not only linked but born in the same instant.

▪ ▪ ▪

Cosmological inflation is the theory that describes what happened in the first fleeting instant after time itself began. It proposes that a mysterious quantum field, known as the inflaton field, briefly dominated the universe, driving an unimaginably rapid and exponential expansion of space.

Recall that you can think of a quantum field as an invisible medium that fills all space and is capable of fluctuating and transferring energy. The inflaton field is one such field, but with a special role. In less than a trillionth of a trillionth of a trillionth of a second, it overpowered everything else, causing a region that might have started barely a meter across to inflate to more than a hundred times the size of today's observable universe.

In the Cosmological Realm, we discussed cosmic horizons, which define the boundaries of what we can see. One of the most important is the cosmic event horizon, which marks the farthest distance from which light emitted "now" could ever reach us in the infinite future.

During inflation, regions that now appear forever cut off from one another because they are on opposite sides of our visible universe and appear to have *always* been beyond each other's cosmic event horizon were once close enough to interact. They shared energy and information. That's why distant regions of the universe today look so similar, even though they're too far apart to have "shared notes" since the Big Bang. They didn't have to. Inflation tells us they already had before they were yanked apart. The uniformity we see today isn't a coincidence or a puzzle. It's a fossil imprint of this brief, primordial moment of extreme expansion.

Postulating a new quantum field like the inflaton may sound as if we were just making stuff up. But this isn't armchair speculation; it's physics. What makes the inflaton field a scientific hypothesis, rather than a SWAG, is that it leads to testable predictions.

First, we assume the inflaton quantum field existed at time zero and drove the inflationary expansion. From that assumption, the smoothness, large-scale curvature, and detailed statistical shapes of tiny fluctuations in the cosmic microwave background have been calculated. Observations have confirmed each of these to high precision.

We'll dig deeper into those details in the Multiverse Realm chapter. (Why there? Because many of us astro-folks suspect that the inflaton field didn't just fire once and call it a day. It may still be ongoing, spawning entire new universes in a process known as eternal inflation. The idea that this field might never stop . . . well, that has consequences. Big ones.)

But for now, here in the Temporal Realm, we're focused on a narrower but equally profound question: how inflation shaped energy flow in the early universe, and how that flow birthed time itself.

To understand that, we must start where it all began: the Big Bang.

But if you ask me, our universe didn't start with a single bang. I say there were up to ten. Of course, it all depends on what you mean by "bang." I define a bang as *a massive transfer of energy from one quantum field to another*.

Two fundamental processes have shaped the evolution of our universe: (1) the continual stretching of space-time that defines our universe's growth; and (2) the natural tendency of energy to spread out and settle into lower states (a.k.a. entropy). The four known field interactions govern these processes: the strong and weak nuclear interactions and the far-reaching electromagnetic and gravitational interactions.

When I describe the universe's beginning as a series of "bangs," I'm pointing to a cascade of transitions: quantum fields settling into progressively lower energy states and transferring the energy difference into other quantum fields as they do so.

There are different ways to count the bangs, but here's my version:

- Two bangs are associated with inflation, within the first 10^{-36} to 10^{-32} seconds.
- One bang marks the annihilation of matter-antimatter pairs within the first second.
- Two bangs involve the universe becoming "neutral" under the strong and weak nuclear forces. This is primordial hadronization and nucleosynthesis, taking place in the first three minutes.

- Three more bangs occur as the universe becomes electromagnetically neutral in stages: at 18,000 years, 100,000 years, and 380,000 years.
- One bang is associated with matter absorbing the electromagnetic radiation from the universe's first stars.
- And the final bang—the tenth—is the one we're still living through: the Stelliferous Era, where matter becomes gravitationally neutral by forming stars and galaxies, transferring mass energy into light.

Let's walk through each of these, starting at the beginning. Or at least as close as we can get, since we can't actually reach time zero. The physics of that moment is still hidden from us, buried under energy densities too extreme for our current theories. And even if we could describe those conditions, we might never find direct evidence, as subsequent events wiped out the traces. It's like a crime scene that's been bulldozed. But we can get close. Really close. We start at 10^{-36} seconds.

BANG 1: THE INFLATON IGNITES SPACE-TIME

At this moment, the inflaton field dumps energy into space-time. The result? Exponential expansion. The universe doubles in size over and over, with about one hundred thousand doublings in a split second. This expansion sends gravitational waves rippling outward and rapidly dilutes any preexisting energy or particles. Our universe's quantum fields are now essentially empty. The universe has no real particles, although virtual particles are always present, albeit fleetingly.

During this expansion, the temperature of our universe plummets by a factor of 100,000. While the energy densities of most quantum fields crash, the inflaton field's energy stays constant. That's key.

BANG 2: REHEATING

When inflation ends, the inflaton field releases its stored potential energy into other quantum fields. This triggers the production of particles, light, and possibly exotic entities, such as magnetic monopoles, axions, or sterile neutrinos. This phase, called reheating, is so energetic that it restores our universe's temperature to what it was *before* inflation began.

BANG 3: PARTICLE-ANTIPARTICLE ANNIHILATION

The particles created during reheating come in pairs: Each particle is accompanied by its antiparticle. When these pairs interact, they annihilate, releasing a flood of photons into the electromagnetic field. Since we start with an equal number of particles and antiparticles, you'd expect a perfect canceling out, leaving a universe empty of matter. But somehow, a tiny asymmetry is introduced: For every billion antimatter particles, there turns out to be a billion and one matter particles. When the annihilation phase is complete, all the antimatter is destroyed, but a small amount of matter remains. That leftover one out of a billion is *everything* you see now. All stars, planets, oceans, and organisms, including you, are made from that small amount of residual matter left over after annihilation.

Imagine a cosmic dance floor at the end of time's first wild party. Billions of dancers arrive in perfect pairs—one in white (matter), one in black (antimatter). As the music crescendoes, each pair meets in a fiery spin, colliding and vanishing in a burst of light, like ballroom fireworks. It's choreographed annihilation. For every white-suited dancer, there's a black-suited partner, so you'd expect the floor to end up empty.

But just before the final song ends, a subtle mistake in the guest list reveals itself. Out of every billion couples, there's one extra white-suited dancer with no partner. When the music stops and the smoke clears, all the couples have vanished in their mutual blaze. But those lonely, unmatched dancers remain.

That one-in-a-billion survivor becomes the raw material for everything. Galaxies, stars, oceans, neurons, poetry, skyscrapers, and you—all built from the leftover wallflowers of the universe's first great rave.

BANG 4: HADRONIZATION AND BARYOGENESIS (PROTON AND NEUTRON FORMATION)

At the end of the annihilation phase, quarks possess so much energy that the bonds that usually combine them into tethered combinations of two quarks (mesons) and three quarks (baryons, such as protons and neutrons) cannot confine them. This changes when our universe plummets to a temperature of 1 trillion degrees Kelvin at an age of 0.00001 (10^{-5}) seconds. First, mesons form in abundance, shortly followed by baryons. Each bond formed sends energy from the matter fields into the electromagnetic quantum field.

By the time our universe is one second old, the two-quark mesons have all decayed (creating even more photons), and only the baryons remain. At this stage, the universe has roughly equal numbers of protons and neutrons. However, that lasts for only a proverbial minute (in reality, it lasts only several seconds). At one second old, temperatures in the universe are so high that protons and neutrons freely convert into each other at equal rates through weak nuclear processes. As the universe expands and cools, the rate of these conversions drops. Neutrons are slightly more massive than protons, so creating neutrons is a bit more difficult at lower temperatures. Neutrons also decay into protons with a half-life of around fifteen minutes. Soon, there are more protons than neutrons; eventually, the ratio settles at seven protons for every neutron.

At this point, our universe is around ten seconds old, with a temperature of 1 billion degrees Kelvin. A great deal has happened in an incredibly short time. Naturally, one might ask, How do we know any of this? Well, we don't know in the eyewitness sense. But it's much more than a SWAG. Using the known laws of physics, we calculate what should happen, deriving precise proton-to-photon and proton-to-neutron population ratios based on the conditions of our early universe. These ratios set the initial stage for the next act: Big Bang nucleosynthesis (BBN). And because the products of BBN are baked into the universe and measurable today, the physics leading up to it isn't just theoretical; it's testable. Our understanding of what came before is defined and constrained by what remains.

BANG 5: BIG BANG NUCLEOSYNTHESIS (SMALL NUCLEI FORMATION)

The term "nucleosynthesis" refers to the creation of nuclei. BBN begins when our universe is around two minutes old and lasts for just under twenty minutes. (We know this because we have re-created similar conditions in experiments.) During this time, temperatures and pressures of our universe's matter and radiation mimic those inside stellar cores, where protons collide, tunnel, fuse, and transmute.

The first step is the combination of protons and neutrons to form heavy hydrogen, a.k.a. deuterium. However, since photons vastly outnumber regular matter, the deuterium nuclei that form are immediately torn asunder by energetic photons. But as the universe continues to cool, this so-called photodissociation becomes less prevalent, allowing deuterium to survive.

Deuterium nuclei then act as seeds for further nuclear reactions, forming the nuclei of ^{3}He (helium with two protons and one neutron) and ^{4}He (helium with two protons and two neutrons). Some deuterium nuclei also form tritium nuclei, which quickly decay into ^{3}He nuclei. Additionally, some ^{3}He nuclei collide to produce ^{4}He nuclei and a proton. ^{3}He and ^{4}He further react to produce ^{7}Li nuclei (lithium with three protons and four neutrons) and two leftover protons.

We know all this because the relics of BBN can be seen today in the primordial abundances of hydrogen, helium, and lithium. The regular baryonic matter in today's universe (to be clear, this does not include dark matter!) is, by mass, roughly 75 percent hydrogen and 24 percent ^{4}He, with trace amounts of deuterium, ^{3}He, and ^{7}Li. (Roughly three-quarters of all hydrogen is primordial, meaning it came into being during the BBN. This means that roughly three-quarters of the hydrogen found on Earth—in water, for example, and in you—is about 13.8 billion years old. We are all children of creation.)

This high primordial abundance of ^{4}He is consistent with the significant conversion of protons and neutrons into helium during BBN, because it cannot be explained by nuclear fusion inside stars. Additionally, ^{4}He nuclei are incredibly stable, meaning the ratio observed today closely resembles that at the end of BBN. Not incidentally, these nuclei are also known as alpha particles—they are the OG particles!

The precise ratio of hydrogen to deuterium is crucial. Because deuterium is easily destroyed in stars by energetic photons (just as it was in the early universe), its observed abundance provides a lower limit on its original abundance. Observations of this ratio in the universe's oldest, most distant gas clouds provide a critical test of Big Bang cosmology. The observed values align well with the predictions from BBN, given the known baryon-to-photon ratio. This is a perfect example of observation confirming a scientific calculation.

Astronomers observe consistent ratios of these elemental isotopes across our universe because our early universe is homogeneous and isotropic on large scales. (As a reminder, homogeneous means our early universe has the same composition everywhere, and isotropic means there are no special directions.) No matter where you look, everything appears roughly the same, and the rules of physics hold at any given point. As a result, the BBN process takes place at all locations in our universe more or less at the same time. Differences in nucleosynthesis would indicate regions of our universe that were significantly hotter or denser than others. However, observations of the CMB confirm a very uniform early universe.

The fact that the observed isotopic ratios of hydrogen, helium, and lithium are aligned with predictions from the Big Bang theory is a compelling argument for the Big Bang's validity. What we see is what we calculated! Pretty impressive if you ask me.

BANGS 6–8: RECOMBINATION
(FORMATION OF FIRST ATOMS)

Now our universe turns its attention to electromagnetism. When recombination begins, the universe is a plasma of small elemental nuclei, electrons, and photons interacting under the influence of electromagnetism. Since the neutrinos have no electric charge and virtually no mass, they do not participate. Instead of being as hot as the core of a star, as at the start of BBN, our universe's temperature is now cooler and similar to that of a star's surface.

The recombination epoch occurs in three steps. First, ^4He nuclei throughout the universe begin to bond with a single electron ($^4He^{+1}$); this occurs when the universe reaches an age of about 18,000 years. These $^4He^{+1}$ ions then begin bonding with a second electron to form electrically neutral atoms; this process occurs when the universe is about 100,000 years old. The final step is when protons—that is, hydrogen nuclei—bond with an electron. The photons released in this step are what we now detect as the cosmic microwave background, our oldest observable light. This happens when our universe is around 380,000 years old.

At this point, the universe's matter is no longer an opaque plasma. It has now transitioned to being a transparent gas of hydrogen and helium. Our universe is still quite hot, at about 3,000° Kelvin, and thus each atom formed possesses considerable kinetic energy. Hydrogen and helium are bouncing all over the place. The process of our magical electrons capturing protons and all those light element isotopes converts their electrical potential energies into photons. This means that the three steps of early universe recombination all represent energy transfer from quantum fields associated with matter to the electromagnetic quantum field. Now, at long last, our universe has largely minimized its energy under electromagnetic interaction. But this won't last.

BANG 9: THE EPOCH OF REIONIZATION (MATTER SUCKS BACK)

After recombination, the universe is filled with a vast, neutral fog of hydrogen and helium atoms. Photons from the cosmic microwave background are free to coast through space, but new photons—from stars and galaxies forming in the darkness—don't get far. Neutral atoms are opaque to high-energy light, absorbing it as fast as it is emitted.

The high-energy photons blast into neutral hydrogen and helium atoms, ionizing them and once again separating electrons from the nuclei they had captured, thereby re-creating the plasma state that had dominated the early universe. This is the *epoch of reionization*, marking another major shift in energy flow.

For the first time, there is a net energy flow from the electromagnetic field back into the matter quantum fields. It is a full reversal in the direction of energy flow. Photons from stars and matter near black holes are transferring their energy into electrons and protons, re-exciting them, tearing them apart, changing their state. Our universe is being ionized by its own starlight.

Reionization is not instantaneous. It begins around 150 to 400 million years after the Big Bang and concludes around 1.1 billion years after the Big Bang. As galaxies form, they create expanding regions of ionized gas around them. Over time, these regions overlap and grow, clearing the remaining fog. By 1 billion years, our universe has become largely a plasma once again.

BANG 10: THE STELLIFEROUS ERA (GRAVITY'S TURN)

One major interaction remains in this cosmic sequence: gravity. While our universe was neutralizing under electromagnetic forces during the recombination epoch, gravity was already hard at work in the background, quietly preparing to reshape everything.

This is where we must reintroduce a key player we've largely left out of the early narrative: our cosmic ride or die, dark matter.

We don't know exactly where dark matter comes from. That mystery is still unresolved. However, all indications, from observations to theory, suggest that it was already present before Big Bang nucleosynthesis began. In fact, dark matter began clumping under gravity as early as ten seconds after the first bang.

As explained in the Dark Realm, our universe's energy content can be broken down into three major types: radiation, matter, and the energy of space-time itself. Matter, in turn, includes both the baryonic matter that makes up stars and planets and the far more abundant dark matter that dominates the gravitational landscape.

During our universe's first fifty thousand years, radiation ruled. Its pressure was so intense that it resisted the collapse of ordinary (baryonic) matter into dense structures. But dark matter, which doesn't interact with light or experience radiation pressure, was immune to this resistance. So while photons kept regular matter diffuse and unstructured, dark matter was already clustering. It was laying down the first gravitational blueprints for what would eventually become the large-scale structure of the universe.

At around fifty thousand years, the energy density of matter caught up with that of radiation. This tipping point, known as matter-radiation equality, marked a fundamental shift. From that moment forward, gravity's influence steadily grew as matter became the dominant energy density. Dark matter structures, seeded earlier, began collapsing more rapidly, creating deep gravity wells. Ordinary baryonic matter, finally free from the dominance of radiation pressure, flowed into those gravity wells. But this was not a one-way trip.

At this time, the baryonic matter is still ionized. Consequently, it interacts strongly with light. As the matter falls into the gravity wells, the high-density photons exert increasing radiation pressure pushing outward. The denser the matter becomes, the stronger the radiation pushes out. This tug-of-war between gravitational pull and the push of radiation pressure results in oscillations in the primordial plasma. Similar to sound waves, these are the acoustic oscillations described earlier in the Dark Realm.

What's remarkable is how this gravitational structuring represents yet another bang: a slow but powerful energy transfer from gravitational potential energy and mass energy to electromagnetic radiation. As matter collapses and heats, it emits light. Every star you see in the night sky is evidence of this transformation. Gravity squeezes, and the cosmos shines.

This process is still going on. We're living in it. The Stelliferous Era is the long, slow burn of matter fields transferring energy into the electromagnetic field, again and again, star after star, galaxy after galaxy.

And according to our best models, this bang will continue for another one hundred trillion years.

What does all of this mean for the Temporal Realm? For one, our universe is young. A lot has happened, but we are still at the beginning of time. Comparing the current age of the universe, approximately 13.8 billion years, with 100 trillion years is akin to comparing a baby that is 3 days and 14 hours old with a person 100 years old. Our universe hasn't yet grown its first tooth, taken its first step, or been potty trained.

Another way we know our universe is young is that *only a young universe is observable*. The galaxies in our local galaxy cluster (the Local Group) are gravitationally bound and not subject to the ongoing acceleration of our universe's expansion. These galaxies will remain visible indefinitely because the gravitational attraction within our cluster overcomes the effects of cosmic expansion.

Eventually, the Local Group will merge into a single massive elliptical galaxy. And then, due to space-time expansion, this single galaxy will become isolated from the rest of our universe's galaxies. This will play out in every galaxy cluster, everywhere in our universe. The time it takes for this isolation to occur, when galaxies outside our cluster recede beyond our cosmic horizon, is "only" around one hundred billion years from now. (In our universe lifespan model from two paragraphs ago, one hundred billion years is the equivalent of a five-week-old infant.) Once other galaxies disappear, our massive elliptical galaxy will become the entirety of our observable universe.

Even before galaxies are completely lost beyond the horizon, their light becomes increasingly redshifted, dim, and faint, making them undetectable with current or foreseeable telescopes. This means the practical loss of visibility occurs much sooner than the absolute isolation caused by space-time expansion. Cosmology, as we currently practice it, will be practically impossible.

Another takeaway from our timeline is that our universe of

matter and light *will* ultimately come to an end, but not for a while. Somewhere between one hundred billion and one trillion years from now, just as galaxy clusters become isolated by cosmic expansion, star formation will cease; all the gas will have been consumed in stars or ejected by stellar winds, supernovas, and accretion disks. Of the existing stars, long-lived red dwarfs will burn out last, enduring for up to ten trillion years. At one hundred trillion years, the universe will be populated by galaxies of stellar remnants (white dwarfs, neutron stars, and black holes) orbiting supermassive black holes. By ten quadrillion years, each galaxy cluster will have collapsed into one or two ultra-supermassive black holes. It will take ten quintillion years (a 1 followed by 19 zeros) for the surrounding space to become so cold that black holes *begin* to lose their mass via Hawking radiation.

As our universe evolves, light is emitted in five main bursts: the matter-antimatter annihilation in the early universe, nucleo-synthesis, recombination, the stelliferous age, and the era of black hole evaporation. However, all this light will have its energy stretched out by cosmic expansion. Even after our universe has transferred all energy from the quantum fields of matter (outside black holes) into the electromagnetic field, cosmic expansion will continue to stretch and dilute that energy across space.

By ten trillion years, around the time our last red dwarfs are burning out, cosmic light will become undetectable. By 10^{68} years, any remaining traces of electromagnetic energy, such as those from stellar remnants and evaporating black holes, will be effectively reduced to zero. This end point is commonly referred to as the universe's heat death.

Another takeaway regards dark matter. There is no accepted story for its origin. It does not help that we have no idea what it is, whether it actually exists, or if it is an undiscovered quirk of gravity. In 2013, scientists hypothesized a Dark Big Bang that would have occurred after inflation but before the acoustic oscillations marking the beginning of structure formation. A new quantum field would be required for this scenario, as with axions and the inflaton field.

Currently, dark matter and dark energy are best understood not as fixed discoveries but as proxies for unknown physics. They're stand-ins for behaviors we observe that current theories can't explain without invoking invisible ghostly stuff. Whether these are signs of undiscovered particles, new quantum fields, or a fundamental misunderstanding of gravity itself remains unknown. In this book, I treat them as diagnostic clues, not final answers. They may be temporary scaffolding in our models, or the first glimpse of deeper foundational truths.

A final takeaway is related to our universe's beginning. What seems clear is that the energy that was and is being converted into our universe's matter, radiation, and space-time existed before our universe was born. It came from the time before time.

I don't blame you if you find this statement nonsensical. Most of us have been taught that asking what existed before our universe is similar to asking what is north of the North Pole. Time exists only *in* our universe, we are told. But in this case, I feel the question does make sense. And here is where I dare to diverge from physics orthodoxy.

As I stated in the Quantum Realm, the existence of energy attests to time's existence. Just as momentum (that is, motion) requires the existence of geometry (that is, space), the fact that energy exists means time exists. Energy generates change. Change requires a difference in the temporal dimension. If energy exists in the time before time, then, a bit illogically, time also exists in the time before our universe's time.

Here, where the Temporal Realm brushes up against the Quantum Realm, things get truly strange. Time itself is subject to quantum fluctuations, just like space. In those earliest instants, before the arrow of time was "burned in," time may have been just as likely to run backward as forward. Physicists suspect that the arrow was set in the forward position when the universe reached the age of a single Planck time, approximately 10^{-43} seconds, after the inflationary event. Before that, the universe may have experienced countless temporal reversals and re-reversals on unimaginably short scales, perhaps adding up—under its own quantum reckoning—to what we would call billions of years.

We already have a candidate for the location of the energy that existed in the time before time. It is the inflaton field. The most precise cosmological data are all consistent with the existence of early universe inflation. This leads us to an unexpected consequence of pre-energy and pre-time. The fields that form the foundation of our reality are ever present in every location, and our universe may not be the only one. Indeed, there may be no end to the number of universes out there. We may inhabit an eternal, infinite multiverse where an infinite number of "you" are always reading this and every other book—like my other book, *A Quantum Life: My Unlikely Journey from the Street to the Stars*.

Don't wait! Step right up and get yours today! Time is running out!

Or is it?

MULTIVERSE REALM

It's hard to build models of inflation that don't lead to a multiverse . . . and evidence for inflation will be pushing us in the direction of taking [the idea of a] multiverse seriously.

—Alan H. Guth

Ideas about the Multiverse Realm represent some serious SWAGs that draw us ever closer to the Realm of Imagination. Diving into the multiverse can feel like opening a book of fairy tales, but unlike fairy tales the strange stories that cosmology and quantum mechanics tell us are backed by mathematics, observation, and experiment. Both the Cosmological and the Quantum Realms, as different as they are from each other, seem to hint at the same intriguing possibility: that our universe is not alone. Not by a long shot.

The idea of a multiverse challenges our very understanding of reality. It's one thing to argue that basic life is prevalent throughout the cosmos or that other intelligent life is all but certain (even if we're unlikely ever to meet them). But can we credibly claim there really are other universes out there? Some consider that tinfoil hat territory.

And yet, if our journey through the Nine Realms thus far has taught us anything, it's that our universe doesn't always conform to our expectations. What is clear to us physicists is that humans are on the cusp of a profound shift in cosmological paradigms. As we push the boundaries of observation and knowledge, we find that our universe (or perhaps the multiverse) is richer, stranger, and more wonderful than we ever imagined.

So, what are these hypothesized multiverses, and where are they? It depends on whom you ask. Several prominent physicists and cosmologists have presented their ideas, and popular culture is replete with examples.

Let's start with popular culture. As you might have guessed by my occasional references to Galactus, I love the Marvel Universe, and my love goes to back in the day before anyone had made even a single Marvel movie. I had two personal interactions with Marvel early in life that captivated me and expanded my imagination. The first, at age nine in 1976, occurred due to nearly unfettered access to hundreds of comic books. That year, I moved to my father's rural village in Mississippi and met my cousins. My father had ten siblings, so there were many cousins to go around. Several boys near my age (and even adult men!) had crates of comics. I would sit in their homes for hours, entranced by the adventures of the Silver Surfer, Thor, Galactus, Power Man, Dr. Strange, Magneto, and Professor X and his X-Men.

But it was not until a decade later, when I was a seaman in the U.S. Navy at age eighteen, that I encountered the *Official Handbook of the Marvel Universe*. This set of fifteen comic books provided a detailed, encyclopedic overview of every Marvel character, including their origins, powers, and vulnerabilities and the physics of their existence. Why doesn't Bruce Banner's transformation into the Hulk violate the conservation of mass and energy? The mass is borrowed from a parallel universe, of course!

Marvel's multiverse is infinite. This means that every possibility is realized. There are infinite universes containing life, and there are infinite dead universes. There are universes where you read this book but have the head of a mouse instead of your own. You're an alligator, an ant, a hawk, a worm, and a shrimp all at once.

In the Marvel Universe, multiple dimensions and realities intertwine. Their stories play with the concept of dimensional travel, where characters leap between worlds, discovering realities that mirror, distort, or echo their own. Heroes from one world encounter their doppelgängers or face existential threats from alternate versions of themselves. Think of Peter Parker swinging through New York's skyline, only to meet another Spider-Person from a universe where Miles Morales or Gwen Stacy dons the mask.

But my all-time favorite multiverse storyline in Marvel concerns a powerful, enigmatic figure known as the Beyonder. The Beyonder was a universe unto himself and was millions of times more powerful than Marvel's entire multiverse combined. Originally, the Beyonder existed in the "Beyond-Realm," a dimension of his own making, where he was the sole inhabitant and the embodiment of everything in that realm.

He discovered our universe when a scientific experiment conducted by the supervillain group known as the Molecule Men accidentally created a pinhole from our universe to the Beyond-Realm. This minor breach intrigued the Beyonder, because it was the first time he ever perceived something outside his own dimension. Until that moment, he was not even aware of the existence of other realities or beings. Dude was lonely, but he didn't even know it!

Intensely curious, the Beyonder peered through this pinhole, gaining his first glimpse of our universe. He saw the complexity and diversity of life, the multitude of beings, and the vast differences in power and morality. These observations fascinated him and, like a scientist, sparked a deep desire to understand and experience these new concepts, leading him to venture into our universe and to Earth.

When I first encountered the Beyonder, I couldn't help but think how cool it would be to talk to a superhero who was an entire universe. While this is impossible, since the Beyonder is only a fictional character, I did once get to meet and spend most of a day with a different Marvel superhero: the Black Panther! Okay, not the character per se, but the actor who played him, the late, great Chadwick Boseman. I certainly entered an entirely new universe that day! But that's a story for another time.

The second multiverse idea I enjoy also involves collisions across dimensions. It comes from string theory and its idea of "brane" worlds (short for "membrane"), which posits that our three-dimensional universe is one of many sheets moving through a higher-dimensional "bulk," containing other sheets or "branes." Each brane holds its own universe, perhaps with entirely different physical laws and its own tales of stars, galaxies, and maybe even life. As the branes move through the bulk, they sometimes collide and create new Big Bangs in the interaction, reenergizing and resetting the interacting branes to time zero.

This model has largely been abandoned, but it's still cool as heck to think about. In this dimension of interacting "membrane" universes, we have Big Splats instead of Big Bangs, as the braneworld universes slap together like hands clapping.

The fact that time resets after brane collisions makes the braneworld a type of *cyclic universe*. This refers to a cosmological model where the universe undergoes a series of infinite, self-sustaining cycles. The universe is born, evolves, and ultimately comes to an end. Then it is reborn, grows again, and ends again—repeatedly into infinity.

Cyclic models are meant to resolve the Big Bang's most nonphysical feature: the initial singularity where the laws of physics as we know them break down. By proposing that the universe undergoes endless cycles of expansion and contraction, cyclic models potentially eliminate the need for a singular beginning—the time before time is just, well, more time.

While intriguing, the cyclic universe models are still highly speculative, and as far as science is concerned, they just don't work. They require further empirical evidence and theoretical development to be considered as viable as the more mainstream cosmological models like inflation, or as fun and awesome as the Marvel Universe "model."

Now let's get down to the two multiverse models that physicists take seriously, starting with the Quantum Realm's *many-worlds interpretation* (MWI), which stems from superposition and measurement.

In the Middle Realm, the rules of classical physics reign supreme. A ball thrown into the air arcs gracefully in a trajectory dictated by its energy and momentum, which have clear, definite values. In this realm, at any given point in time, a ball or any other object has a single speed, position, and energy, and you can know all three simultaneously. This one-to-one correspondence between reality and our measurements feels natural because it matches what we experience every day.

But as we know, the Quantum Realm is different; it's *Through the Looking-Glass* territory, where reality is written not in cause-and-effect prose but in seemingly nonsensical statistical poetry. Here, quantum particles, which are better understood as excitations in quantum fields, refuse to sit still for singular descriptions. Instead of having one specific energy, momentum, size, or location, quantum entities are described by their probabilistic wave functions. Recall our flipping coin, which is neither heads nor tails but both *heads and tails*, until you catch and observe it, at least. That's quantum superposition.

This raises a profound question: What happens when we measure a quantum system? In the Middle Realm, measurement is straightforward. You point a speed gun at a moving car, and voilà! You get a number. But in the Quantum Realm, measurement doesn't just reveal reality; it *creates* it. When we measure a quantum particle, whether to find its position, spin, or momentum, the wave function appears to collapse. It's like a magician snapping their fingers and choosing a single rabbit from a hat filled with endless potential rabbits. This collapse suggests that the wave function isn't physical reality but a tool for predicting probabilities. In this view, measurement transforms the fuzzy realm of possibility into the crisp world of certainty.

In contrast, the many-worlds interpretation of the multiverse refuses to let the other possibilities vanish when one possibility is "observed." Instead, it claims they *all* happen, but in separate branches of a larger multiverse. Each time a quantum measurement is made, reality doesn't collapse; it splits, spawning a new universe for each possible outcome. All possible universes are equally real, though forever isolated from each other by quantum multiverse horizons. These parallel universes aren't like neighboring houses on the street but more like separate dimensions. For example, imagine measuring the spin of an electron, which could be either "up" or "down." In the many-worlds view, if you observe it as spin-up, another version of you observes it as spin-down in a different branch. These branches don't interact; they're always present but never touching.

The many-worlds interpretation has a certain elegance but also staggering implications. It suggests that infinite versions of every particle in existence—and every configuration of particles, including you—exist across these branches, each living out different choices and outcomes. In one, you're a world-famous rock and roll rap star. In another, you've won a Nobel Prize in physics. In yet another, you're arguing with your dog over the remote control.

But this flood of universes comes with challenges. For instance, how do we deal with probabilities? In standard quantum mechanics, probabilities predict outcomes. But in MWI, where every outcome occurs somewhere, what does "probability" even mean? If every outcome exists, then there is no probability, no winners, no losers. It's like betting on a horse race where every horse wins, loses, and comes up lame and is sent to the glue factory in some parallel reality.

Then there's the *basis problem:* What determines which events cause the universe to branch? Does the universe split when a photon hits your eye? What about when you decide to skip breakfast or pet your dog? Does this branching happen continuously with every quantum interaction? Could it be that each time an atom jiggles, each time a photon scatters, the universe splinters into a cascade of parallel realities? That results in an ever-growing multiverse filled with endless versions of everything that could happen. Without clear rules, the concept becomes unwieldy. Not to mention, energy conservation is obliterated. In this model, there is an unlimited amount of energy available at all times in every direction.

And of course, there's the sheer complexity of infinite branching. MWI violates the spirit (as well as the letter) of Occam's razor, which favors simpler, more straightforward explanations. Suggesting the universe splits endlessly to account for quantum phenomena can feel like explaining a missing sock by proposing a parallel sock dimension.

I'm not convinced.

The Cosmological Realm also raises the possibility of a multiverse, and here this idea is better supported by the data. While still speculative, they are grounded in physical theory and cosmological observations, making them appear more robust, if no less mind-bending.

To see what the Cosmological Realm has to say about the multiverse, we need to revisit the inflaton field introduced in the previous chapter. This field is central to inflationary cosmology, a concept that explains the universe's rapid expansion in its earliest moments. We take inflation seriously because it has made predictions that align almost perfectly with our observations.

A cornerstone of inflationary theory is its predictions about the cosmic microwave background—the faint glow left over from the Big Bang. Inflation predicts that the CMB's temperature fluctuations will be strikingly uniform across the sky and follow a bell-curve-like distribution. Observations confirm this to astonishing precision.

Even the fine details, such as the "acoustic oscillations" in the CMB, align with inflation theory predictions. These peaks are the ripples caused by sound waves traveling through the hot plasma of our early universe. It's as if we were listening to an ancient cosmic symphony and inflationary cosmology predicted its notes.

And the music has another layer: correlations that stretch across parts of the sky so far apart they should never have been in touch with each other . . . unless inflation happened. When scientists compared the CMB's temperature map with its polarization map, they found patterns lining up across these vast "superhorizon" scales. A simple Big Bang without inflation can't explain such long-distance harmony; only inflation, which ballooned tiny quantum jitters into sky-spanning fluctuations, can do this. The fact that these superhorizon correlations appear exactly where inflation predicted is one of the clearest fingerprints

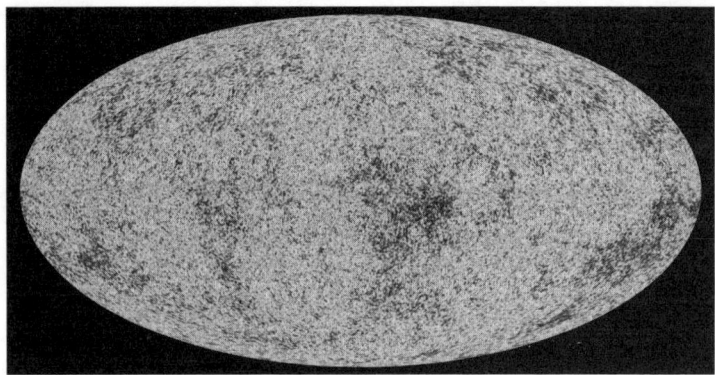

Fig. 20: Planck CMB
Cosmic microwave background temperature fluctuations across angular scales—evidence of inflation and the universe's early structure.

we have that the early universe underwent this dramatic early growth spurt. The time before time is real!

Inflation also explains how galaxies came to be. Quantum fluctuations during inflation acted as "seeds," creating tiny variations in matter's density. Over billions of years, these seeds gravitationally gathered more and more material, growing into the galaxies we see today. Large-scale galaxy surveys reveal a distribution that matches the predictions of the inflationary model. This even extends to the baryonic acoustic oscillations mentioned above—subtle patterns in the arrangement of galaxies that serve as a "fossil record" of sound waves from the early universe.

Perhaps the most exciting test of inflation is the search for primordial gravitational waves, ripples in space-time that were created during inflation. These waves, if found, would serve as a direct signature of inflationary expansion. While we haven't detected them directly, their effects may be evident in the polarization patterns of the CMB.

One specific type of polarization, B-mode polarization, holds the key. Imagine the electric field of light arranged in swirling, circular patterns; this is B-mode polarization.

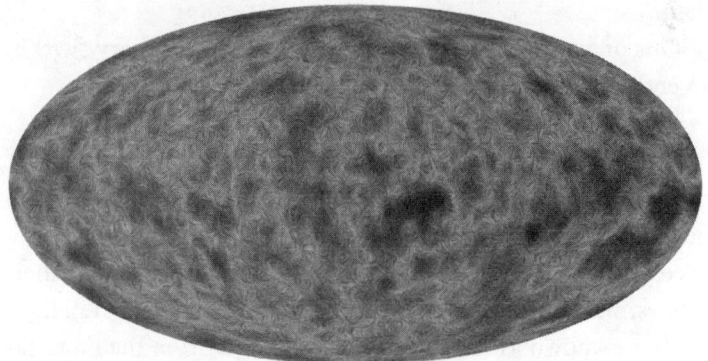

Fig. 21: Planck Polarization Map
Polarization of the CMB, revealing early universe dynamics and supporting the ΛCDM cosmological model.

Detecting these patterns could not only confirm inflation but also reveal the energy scale of our universe's early rapid expansion.

This is where inflationary cosmology ties directly into the multiverse. The theory posits that our universe originated from a rapidly expanding inflaton field, which erased any trace of its previous state. Imagine a one-meter sphere ballooning to the size of the observable universe in an instant, creating a space-time tsunami that wipes everything clean.

But this inflationary process doesn't come to a universal halt. Most of space continues expanding at breakneck speed, driven by the persistent energy of the inflaton field. Yet, due to quantum fluctuations, isolated regions randomly decay. They cool, slow their expansion rate, and drop out of inflation mode. These pockets nucleate into distinct "bubbles," each undergoing its own Big Bang–like evolution.

Inside these bubbles, physics takes root. The field stabilizes, matter forms, forces differentiate, and the familiar laws we measure today emerge. But outside, inflation keeps roaring ahead, spawning more and more such pockets. Each one is causally sealed off from the others by the ever-expanding space between them.

Our universe is just one of these bubbles—a solitary swirl in an endless, churning sea.

Welcome to cosmology's multiverse.

If this is accurate, it forces us to confront the question: What the heck is the inflaton field? It was once thought that the Higgs field, predicted by the Standard Model of particle physics as a necessary mechanism for endowing quantum particles with their masses, was the inflaton field. Both the inflaton and the Higgs fields are known as scalar fields. These differ from the more familiar vector fields, such as electric and magnetic fields, which have a range of magnitudes and directions at any given location.

Scalar fields, by contrast, are described by a single value (a scalar) at each point in space and time, and they don't have any directional properties. Scalar fields are simpler to describe than vector fields, and they span the entire universe. Consider something as everyday as temperature. In any room, a scalar field can be used to describe the temperature at each point. It's 69.7° down here, and 73.2° up there. Those are scalar values of temperature. As a field, temperature has a value everywhere—in the cells of your body, in the air around you, in the center of a star, and in diffuse interstellar nebulas.

But temperature is not a fundamental quantum field, and before 2012 no scalar quantum fields or their associated particles had ever been observed or directly measured. But on July 4 of that year, scientists at CERN's Large Hadron Collider confirmed the existence of the Higgs particle (or boson, as we nerds call it).

This discovery rocked every physicist, but it rocked me to the core. Before 2012, I did not fully accept that any quantum fields—whether scalar or vector—were real entities. I thought it was more likely that they were just mathematical models that physicists created to reproduce and predict reality at the quantum level. Physicists do this all the time; we create models that we can use to make predictions, but they don't necessarily represent the underlying reality. Take the atom. Sometimes we model it as an electron and nucleus connected with a spring; other times we model it as an electron orbiting a nucleus. It all depends on what answer we're looking for. The important thing to note is that we can get the right answers using these models even if they are not reflective of true reality, which in the case of atoms is better described by wave functions and potentials.

We created the fundamental quantum fields, such as the electron quantum field, as a mechanism to address the fact that matter particles pop in and out of existence (these are the pair production and annihilation processes I mentioned in the Quantum Realm). We discovered the particles, observed their behavior, and then created their fields to better understand and work with them mathematically.

With the Higgs, it was the other way around. We knew particles had mass, and the mathematics led us to a mechanism—the Higgs field—that allowed us to account for the creation of mass. This mechanism worked and was useful, but we did not know if the field existed or if there was even such a thing as an actual Higgs mechanism that created mass for all particles.

Cut to July 4, 2012, when scientists found the Higgs particle. When its existence was confirmed, its field's existence was *also* confirmed. The Higgs field was not merely a mathematical concept; it was a physical reality. The irrefutable implication here is that all the other fundamental quantum fields—scalar and vector—that apparently permeate all space can also be real.

Including, perhaps, the inflaton field. (More data are required to fully prove its existence.)

The discovery of the Higgs field placed us on the razor's edge where our collective knowledge rendered us almost godlike. So much had to occur before its discovery was even possible. Our human imagination had to evolve over millions of years to conceive theoretical physics and then suggest the existence of the Higgs field. The profoundly complex Large Hadron Collider, which consists of a particle accelerator measuring seventeen miles in circumference and two primary detectors, ATLAS and CMS, had to be thought up, engineered, built, and calibrated so the collider could slam together beams of protons at energies that exist only in our universe's most extreme natural environments. (Some scientists were even concerned that when it was turned on, the collider would create microscopic black holes that would turn Earth into a giant sinkhole and gravitationally consume itself. That obviously didn't happen—at least in this branch of the multiverse!) The collider's computational infrastructure for storing and transferring data had to be the most sophisticated and expansive in the history of the world up to that point. Finally, we had to learn how to convert the voltages and bits produced by the complex detectors into mathematical statements of measurement accuracy, uncertainty, and confidence.

When you put it all together, this remarkable, near-impossible tool has revealed the proton's inner structure to unprecedented detail, allowing us to measure the masses of nuclear particles, such as quarks, W bosons, and the Higgs particle, to within fractions of a percent.

It does not always work out the way it has for the Higgs particle and its field; the history of science is littered with many more failed ideas than confirmed ones. That it works at all, especially in these more ephemeral and esoteric corners of our cosmos, is ridiculous. And that is a vast understatement. As Einstein once wrote, "One should expect a chaotic world, which cannot be grasped by the mind in any way. . . . The fact that it is comprehensible is a miracle."

Due to measurements of the Higgs field and theoretical investigations of the inflaton field's properties, we now know that these two fields are not the same. Yet the Higgs field, the inflaton field, and the field associated with dark energy are all scalar quantum fields that affect reality at the most profound and fundamental levels. We haven't directly observed particles associated with inflation or dark energy, unlike the Higgs particle. Still, inflationary cosmology offers our most compelling framework for understanding our universe's origins and hints at— some physicists would say *necessitates*—the existence of an infinite multiverse.

Is this still science if we can't ever see those other universes? That's the big knock against the multiverse: There may be no way to falsify it. And that's fair criticism. Science, at its core, is supposed to be testable. But here's the thing about scientific theory at the frontier: Many great ideas start as *testable later*. Not every equation that Einstein scribbled had a clear path to the lab. Take black holes. General relativity suggested they must exist, yet Einstein was reluctant to accept that nature would create something as weird and topsy-turvy as a black hole. During his time, there was no way to unambiguously observe them. But then, decades later, after many technological advancements, we did observe and test them.

The value of these kinds of avant-garde ideas isn't just in what they can explain today. It's in the doors they open for tomorrow. A theory that isn't yet falsifiable can still help us frame better questions, build new tools, and expand our imagination responsibly. That's the tightrope walk of speculative physics. It's not about blind belief. It's about informed curiosity.

The challenge lies in testing these ideas. While we've made progress through our observations, viewing an infinite multiverse with one's own eyes requires a godlike, multiverse-spanning perspective, like the one the Beyonder had when he stepped through that pinhole and viewed our universe in all its wonder, variation, and complexity. While our sophisticated observational tools help us see what we cannot with our own eyes, our imaginations can see further, even into realms beyond our horizons, where our instruments can never reach.

REALMS BEYOND HORIZONS

The horizon of many people is a circle with a radius of
zero. They call this their point of view.

—Albert Einstein

Imagine being tinier than an atom. Subatomic you is so small
that, for you, a human cell is as big as the Milky Way is to
regular-old, super-sexy human you.

Take a moment to look outward and explore your subatomic-
and cellular-sized universe, just as humans have studied galaxies
to the limits of our observable universe. First, you may notice
that cells come in different types. You see nerve, muscle, bone,
and blood cells, and devise ways to tell these apart from one another. Over time, you sharpen your observational techniques
and learn to see the components inside these cells: ribosomes,
mitochondria, cytoplasm, and nuclei. Then, increasing your
resolution and utilizing precision techniques, you observe the
more minor constituents of these substructures: proteins, fats,
and sugars. Moving in the other direction, you widen your gaze
and learn how cells organize into superstructures that work together in complex and sometimes mysterious ways.

But after performing and cataloging all of these observations,
you invariably come up against an observational horizon that
you cannot get past. No matter how hard you look, you still have
no idea, and no hope of ever discovering, whether you are inside

a whale, a dinosaur, or a tree. You don't even know what any of these words mean, since you never get to see your "universe" from the outside, to say nothing of the ecosystem your "universe" lives in—an ocean, a grassland, a jungle.

We regular-old, super-sexy, everyday humans face this same conundrum as we observe and try to understand our universe. We can't directly observe the *context* of the observable universe; therefore, we can never know some fundamental facts about our universe's existence. We may not even possess the language to describe it.

We will never be able to get outside the universe and view it from a distance. Put a little differently, we will never know what we are a part of or even if we are a part of anything.

Since the earliest days of human history, we have gazed up at the heavens, striving to understand the cosmos and our place within it. Over time, we have uncovered the mysteries of our planetary neighbors and their moons, discovered our place in the Milky Way, peered deeply into nearby galaxies, and even glimpsed those at the edges of time and space. We have deciphered the inner workings of stars and the galaxies they inhabit, revealing how these galaxies are organized into vast structures of clusters and superclusters. These clusters are anchored within a massive framework of dark matter filaments, which are gravitationally collapsing, even though they are surrounded by vast voids dominated by dark energy that drive our universe's inexorable expansion.

But what is our universe's bigger story? What lies beyond our "event horizons" of space and time?

We typically associate event horizons with black holes: Once something crosses a black hole's event horizon, its fate is sealed. No matter what it is—light, matter, dark matter—once inside, it will never return.

But as we've learned, event horizons aren't just for black holes. Generally, an event horizon is a space-time boundary across which information can travel in *only* one direction: away from an observer. (In this case, away from us.) Beyond this event horizon, information can never be sent back in across it. The event horizon of space and time marks the absolute boundary of what we can know.

Event horizons are some of the most extreme phenomena in our cosmos. These mysterious boundaries form where space-time curvature reaches critical intensity or where space-time expansion exceeds the speed of light.

As a reminder, our universe is expanding, and the expansion rate increases linearly with distance. A galaxy twice as far away moves away twice as fast; one three times as far away recedes three times faster. At a certain distance—about fourteen billion light-years away—the recession speed reaches the speed of light. This marks the Hubble sphere, a sort of intermediate horizon.

Our cosmic event horizon lies about two billion light-years *beyond* the Hubble sphere. We can still see galaxies in this zone, which are receding faster than the speed of light, because the Hubble sphere itself is expanding, overtaking photons emitted from galaxies beyond it. Over time, these photons enter a region where space-time expands slower than the speed of light, eventually allowing them to reach us. It's like layers of nested and ever-expanding bubbles, with photons traveling inward from outer layers until they reach our instruments.

This interplay of horizons defines the limits of our observable universe and highlights the complexity of studying the universe on cosmological scales.

The reality of these cosmic horizons is even more nuanced than what I've outlined so far. While I focused on the horizon for signals emitted today, there are galaxies much farther away—nearly three times beyond the cosmic event horizon—whose light we are receiving right now. This farthest observable boundary is known as the particle horizon, located *forty-six billion* light-years away.

But how can we see light that should take 46 billion light-years to reach us if the universe is only 13.8 billion years old? This is because the light has not traveled for 46 billion years, but rather the object that emitted it is now 46 billion light-years away. As the light traveled to us, the galaxy that emitted it was carried rapidly away by our universe's expansion. The light from these galaxies originated or passed through a region of space-time that was within our cosmic event horizon, which is also expanding, allowing this light to reach us today. Any light emitted from these galaxies *now* will never reach us (and vice versa).

The light we see today from objects well beyond our cosmic event horizon tells us something remarkable: The universe beyond appears to be more of the same. The observations suggest a vast, homogeneous, and isotropic cosmos—a seamlessly expanding web of galaxies embedded in space-time, extending into what might seem like the infinite expanse of "nothing."

Have you ever seen one of those end-of-show zoom-out sequences, like in the movie *Men in Black* or the TV shows *The Simpsons* and *Family Guy*, where we discover that our universe is a tiny component in some higher dimension with its own planetary surfaces and sentient life-forms, or where we're contained within a bauble on a cat's collar? Well, the data don't support that. We're not attached to some cat somewhere. I agree it would be much cooler to be on a cat's collar, but that ain't what's up!

How do we know this? Because of our unique observational ability (unique on Earth at any rate). The variations and distributions of temperatures in the cosmic microwave background are consistent with the distribution of matter in the early universe, which remains consistent to this day.* *When* you look matters but not *where* you look. At each when, the wheres are practically the same. It is what it is, no matter where you look at it: super-cosmic goo.

---

* Disclaimer for nerds: the Hubble tension notwithstanding.

But if there were a higher context where things were not so uniform, could we discover this fact by observing what we can and do see? Believe it or not: maybe.

Let's revisit the analogy at the beginning of this chapter, where subatomic you was so small that a living cell would be as big as the Milky Way is to regular-human you. Let's calculate the size of the corresponding observable universe in this scenario. Since the observable universe is six orders of magnitude larger than the Milky Way, the analogous universe would be six orders larger than a ten-micron cell. That is ten meters.

In our analogy, imagine such universes inside a thirty-meter-long blue whale. One observer in one universe could be in its brain, another in its gut, and another way back in its tail. Each has a ten-meter horizon, and none of these horizons overlap. Each observer is checking out what's around it, but they will never observe each other and they will never realize they are all part of a blue whale.

All three observers see their "universe" as composed of cells possessing common substructures. The superstructures are where the differences show up. The end section will have connections leading to fins and a tail. The brain observer will find their universe completely surrounded by bone. The gut will consist of tubes with dynamic interiors changing on much faster timescales than in the other environments. If the observers compared their observations, which they cannot, they might realize that homogeneity and isotropy break down at sufficiently large distances.

Observed from the outside by a human of standard size, it is obviously a whale. But our subatomic-you-sized observers could be fooled into thinking their universe's composition and structure continue to infinite distances. It's as I said at the book's opening: Size matters.

Every location in the universe has its own cosmic event horizon. When the separation between two points is more than twice the radius of their respective observable universes, these cosmic event horizon bubbles cannot overlap. Each observer at the center of such bubbles is forever isolated from the others. They can never exchange signals to communicate. Our larger contextual universe potentially contains an infinity of such isolated spherical sub-volumes.

What can we say about the universe beyond our particle horizon? Regarding its size, inflationary cosmology suggests that our entire universe is at least a couple of hundred times *bigger* than the observable universe. This difference is roughly equivalent to comparing the volume of Earth with that of a body twice the size of the Sun (recall that a million Earths can fit inside the Sun).

Several cosmologists propound even larger versions of our universe, so big that its size must be written as ten to a large power, to a large power, to a large power. Otherwise, there is no practical way of writing it down. Indeed, even a universe infinite in extent is not inconsistent with cosmological theory or data.

As I described in the Temporal Realm, we can tease out the size of the volume that constitutes our *observable* universe, along with its average temperature, component densities, and particle ratios across virtually all time. However, nothing is known that constrains the volume of our *entire* universe, wherever the heck that happens to end.

There are rumblings of the potential to gain information about the universe beyond the particle horizon by observing signals in the CMB or long-wavelength gravitational waves. But what's really needed is the ability to jump from the head of the whale down to its tail. And I mean, physically go there. Strangely enough, the universe provides us with a possible mechanism to achieve such a feat. All we need to do is cross to the other side of *just* the right horizon, where the Cosmological Realm and the Quantum Realm intersect.

We need to cross the horizon of a black hole.

▪ ▪ ▪

Black holes are often described as cosmic vacuums—empty pits in space where everything, even light, gets swallowed into oblivion. But that popular picture is misleading. It's almost like describing the ocean as a giant hole in the ground without mentioning the water inside. The truth about black holes is that they may be much stranger and far more fascinating: They may not be empty at all. In fact, according to a concept called "frozen stars," they're overflowing, packed with the matter that created them in the first place.

Think of the conventional black hole anatomy: a spherical event horizon marking the point of no return and a point-like singularity lurking invisibly at the center, with only a highly curved space-time void in between. But of the multitudes of black holes scattered across the Milky Way—hundreds of millions, at least—almost none fit this minimalist description. Why? Because every stellar black hole is not empty but full of the time-frozen remnants of its own making.

Consider what happens when an observer watches an object fall into a black hole from afar. As the object approaches the event horizon, space-time bends so intensely that time slows to a crawl. From an outside observer's perspective, the falling object seems to freeze, forever hovering just above the event horizon.

Now apply this same thinking to a massive star—say ten times the mass of our Sun—collapsing into a black hole. When such a star's core runs out of nuclear fuel, the star loses the outward radiation pressure supporting its outer layer, causing the star to collapse violently inward. It is now destined to collapse into a black hole. To an outside observer, however, just like the object above, the surface of the collapsing star appears to freeze in time the moment it reaches the event horizon that its collapse just created. This is not just an appearance. Time really does move at a different rate.

This means that lurking beneath the event horizon of every stellar black hole lies the ghostly surface of the original star that created it, still collapsing, still present. Stellar black holes are not hollow, empty places, but spheres densely packed with matter and light. These are called black hole mimickers.

Now imagine you are the object falling into one of these. As you hurtle through the darkness of interstellar space, you might not even notice the black hole ahead of you; it's just a pinprick, maybe twenty miles across for a black hole that is ten times the mass of the Sun. But then, as you cross the event horizon, something extraordinary happens. One moment, you're in the blackness of space, and the next you're on the surface of a collapsing star.

But this surface isn't what you'd expect. Inside a black hole, the rules of physics warp beyond recognition. Everything here moves in only one direction—toward the central singularity. Light can't escape, and heat can't radiate outward. The collapsing stellar material is still there, and it's millions of degrees. But you can't feel it or even see it, even though you're right on top of it. That's because inside the black hole no information, including light or heat, can move "upward" toward the event horizon. As a result, you could never see anything interior to your position.

This is probably a good point to address one of the questions I have been asked most over the years: "Do we live inside a black hole?" At first glance, it sounds wild. But strangely, if you do a simple back-of-the-envelope calculation, it looks as if it could be true.

Here's the basic thought: A black hole forms when you pack a lot of mass into a small enough space that not even light can escape. There's a critical size limit—called the Schwarzschild radius—below which any mass becomes a black hole. If you shove too much matter into too small a space, gravity takes over completely, and boom: black hole.

Now let's apply that same idea to our universe. If we calculate the total mass and size of our observable universe and plug those numbers into the black hole formula, we find that the size of our observable universe is approximately what you'd expect if the universe were a black hole.

But here's the thing. You can't do the Cupid Shuffle inside a black hole! There's no back and forth, left and right, or Charlie Brown inside a black hole! The very fact that you can push a kid on a swing is a conclusive experiment. Where it comes to the inside of a black hole, they not like us.

Let's say you fell into the black hole today. Millions of years later, I fall into the same black hole. Even though I arrive long after you in *outside-the-black-hole* time, I'd find myself on the very same collapsing stellar surface, hurtling inward alongside you in *inside-the-black-hole* time. Time, as we know it, stretches dramatically inside a black hole. From an external perspective, every bit of matter that ever fell into that black hole across millennia appears frozen in space just above the event horizon, yet from your vantage point inside, all of it collapses to the core simultaneously.

From the outside, a black hole looks like a sphere with a radius. Using basic geometry, we can calculate its volume just as we would for a soap bubble floating in the air. But on the inside, a black hole is anything but spherical. Its volume doesn't exist statically; it grows, expanding with time as if the universe were conspiring to make room for its inflow of contents. TARDIS, anyone?

From the perspective of an outside observer, a star collapsing into a black hole appears to be a process without end. Time slows to a crawl at the event horizon, stretching moments toward infinity. What looks like a perpetual pause might, in a sense, be nature's way of sidestepping the problem of singularities, those troublesome infinities that our equations insist should exist. Even after a hundred trillion years, the star's implosion is still, from an external viewpoint, unresolved.

If this isn't strange enough, the same physics that gave us black holes also teases the existence of white holes and wormholes.

White holes are the time-reversed counterparts of black holes, where matter and energy have no choice but to move in the other direction: outward. Wormholes might serve as space-bending bridges to other parts of space-time. The key to these possibilities lies in "negative" quantities: negative distance, negative energy, and negative time. Outside a black hole, these ideas are absurd, akin to asking someone to walk a "negative mile" or age in reverse. However, within a black hole's event horizon, such concepts are legit.

Wormholes are a particularly enthralling concept. It's how we might jump from one end of our cosmic whale to the other (so in this case, it would be more like a whalehole). Picture them as tunnels with a black hole at one end and a white hole at the other. These two mouths are connected by a "throat" that isn't a location in space-time as we know it but a region of negative distance. Imagine two particles simultaneously falling into a black hole from opposite sides. In normal space-time, you'd expect them to meet at the center. But in the bizarre topology of a wormhole, they bypass each other entirely, slipping through the throat to emerge elsewhere or elsewhen.

Early calculations suggested that these wormholes, though mathematically possible, would snap shut instantly, making them useless for travel. But recent developments hint that with the right kind of negative energy—something already observed at the quantum level—it might be possible to prop a wormhole's throat open long enough for traversal. In this scenario, you could jump into a black hole and emerge from a white hole billions of light-years away—or perhaps in a different universe altogether!

Of course, there's a catch. As much as wormholes inspire sci-fi dreams, there's no direct evidence they actually exist. Our best data on what happens to stuff when it falls into a black hole come from black hole mergers observed through gravitational waves. These collisions leave no room for matter or energy to slip into alternate dimensions; everything is accounted for, and nothing vanishes. But these observations involve "young" black holes—stellar-mass objects formed from dying stars. The older, more enigmatic black holes—the supermassive ones lurking at galactic centers—remain an untapped resource in gravitational wave science.

Supermassive black holes are ancient relics, their origins so entwined with the early history of our cosmos that we can't say whether they birthed galaxies or the other way around. Surrounding these colossal chicken-and-egg objects are stellar systems older than almost anything else in the universe. Take the stars in the Milky Way's central bulge: On average, they are twice as ancient as those in its disk. When we look in the other direction, at the stars in the Milky Way's surrounding halo, we find globular clusters containing stars that are so old they once seemed older than our universe itself, a paradox resolved only by refining our cosmological measurements.

And yet black holes are not mere puzzles for physicists to toy with. They are central to the story of existence. Without them, the cosmic chain of events leading to us would unravel. To have life, we need planets, which require stars, which form from galactic nurseries, which seem to rely on the stabilizing influence of ancient supermassive black holes. A universe without black holes would be a universe without us.

In the distant future, as the cosmos continues its relentless expansion, black holes may become the universe's final monarchs. Every star, every planet, and every trace of matter and its energy will eventually find its way into their clutches. But even as black holes dominate the landscape of a dark-energy-driven universe where expansion reigns, what happens deep within their event horizons remains shrouded in mystery. Inside these realms of warped space-time, our physics reaches its limits, leaving us with only tantalizing fragments of insight into the ultimate fate of everything we know.

For the time being, we are stuck at one end of the whale. Whether that fills you with hope or dread depends on your personal philosophy, I suppose.

At this intersection of hard science and philosophy, we reach our final destination, a land that is simultaneously known to us and completely foreign, a place that is almost boring in its regularity but that never ceases to surprise. How it works remains a mystery, even though it is with us for every moment of our respective existences.

Welcome home to the Realm of Imagination.

REALM OF IMAGINATION

Before we go, can you imagine?
Picture years, with your mind, can you imagine?
Paint a picture in the sky, can you imagine?
Call emergency, I've been dreamin' all my life, all my life.

—D'Angelo

Man, I ain't afraid of the universe. I'm *from* the universe!

—Me

In the early chapters of this book, I mentioned that life, while likely prevalent throughout our cosmos, will not always exist. There are few places in our universe where life can be both protected and nourished. And, as we have learned, there is a *lot* of time left before our universe comes to its end. Trillions upon trillions of years from now, our universe will succeed in its ultimate mission to destroy all matter, living or not. The stars suitable for hosting planets with multicellular life will all die. We are the only species on our planet aware of this.

Here is where human minds can showcase their greatest strength: I believe that in humanity's deep future we will overcome these challenges as they exist in our solar system, thereby ensuring life's continued existence beyond the expiration date of planet Earth, our "pale blue dot."

What might this look like for us? What does the science indicate could happen here?

Before we go any further, let me be clear about what this chapter is and what it is not. This is the Realm of Imagination. Here, I allow myself to move fluidly between the credible and the conceivable. This is not sci-fi masquerading as science. It isn't fan fiction for techno-optimism. But neither is it bound by the caution tape of today's engineering constraints.

The goal isn't to predict the future. It's to widen the angle of our gaze, test what our scientific understanding permits and what our technology might eventually enable. If some of the ideas in this chapter sound speculative, that's because they are. But they're grounded in real physics, real trajectories, and real existential stakes.

I write from a perspective that is scientific yet not reductionist, techno-humanist yet not naive, reverent yet not religious. Human imagination is not just a cultural flourish; it is a force of nature, an evolutionary adaptation as consequential as opposable thumbs or spoken language. If life has a future beyond Earth, it will be because imagination got us there.

Roughly 1.8 billion years from now, the Sun's expansion will crank Earth's temperature high enough to trigger runaway precipitation, which will strip carbon dioxide from the atmosphere. Without CO_2, photosynthesis grinds to a halt, and Earth's food web collapses. At that point, Earth will be about 6.3 billion years old and facing its penultimate mass extinction.

Fast-forward five billion more years, and the Sun, now a red giant three hundred times its current size, will have engulfed Earth. Mars will be left where Mercury is now—uncomfortably close to the solar surface. The inner solar system will be uninhabitable. But don't panic. Life has time. And intelligence buys more time.

Life on Earth has survived apocalypse after apocalypse, thanks to the deep playbook of evolution. The hardiest life-forms don't just adapt; they improvise under pressure, testing survival strategies in parallel. As we saw in the Realm of Life, viruses and prokaryotes are the ultimate survivors—small, simple, and possibly widespread across the solar system thanks to cosmic collisions. In a universe that loves smashing things together, life finds ways to spread.

Humans are unique. Our minds can simulate possible futures—including extinction. More important, we can envision ways to overcome them. We don't have to rely on random luck or cosmic slingshots. We can design and create spacecraft, life-support systems, and entire ecosystems from scratch. Microbes adapt through trial and error. We intentionally edit genomes.

Yes, Earth 2.0 is probably a fantasy. And if we do find it, it'll still be far, far away—even with relativistic travel. But we may not need a twin. We may just need places we can tweak—worlds we can tailor with bioengineering, terraforming, and a whole lot of grit.

That effort will start here, in our own solar system. Mars, Ganymede, Titan, and the Moon are the top candidates for human habitation beyond Earth. Mars is close and familiar. Its days are like ours, and at the equator summer temperatures can hit 68°F. But it's dry and has no radiation shielding, and when the Sun balloons into a red giant, it'll be too damn hot.

Ganymede, our solar system's largest moon, may hide its biggest ocean beneath the moon's icy shell. It's also the only moon with a real magnetic field. But it orbits inside Jupiter's caustic radiation belts and lacks an atmosphere.

Titan offers something different: thick nitrogen air, lakes of hydrocarbons, and surface pressure similar to Earth's. It lacks oxygen, but it's safely beyond Saturn's radiation belts yet still wrapped in its protective magnetosphere. It's cold and alien, but at least it's shielded.

As for our Moon, it'll be mined, studied, maybe settled, but probably by robots more than humans. Think Antarctica meets Rikers Island meets Chernobyl.

If we manage to spread across these worlds, it'll be as astonishing as life's first spark on Earth. If we go farther, to other stars, it'll be miraculous. But maybe we won't need to. Maybe our tech will let us stay in our own solar system. Maybe we'll even learn to survive without the Sun itself. Everything we need might already be right here.

A primary practical challenge for humanity's future is that we must rely on increasingly difficult-to-reach energy sources to power our societies. While we work on planning, inventing, and designing our path outward, we must also take inspiration from the *Star Trek* transporter room and find ways to *energize* ourselves.

In 1964, the Russian astrophysicist Nikolai Kardashev classified three types of civilizations: Type I, Type II, and Type III. To reach Kardashev Type I status, humanity would need to harness roughly 10^{16} watts of power—about what Earth receives from the Sun—effectively tapping all of our planet's available surface energy. So far, we have harnessed only about 0.01 to 0.1 percent of this energy. Translation? We've got a long way to go—and a whole lot of solar panels to build!

To reach Type II civilization status, humanity would need to capture and utilize the full energy output of the Sun. This is usually envisioned through the construction of a Dyson sphere— a theoretical megastructure designed to absorb solar radiation and convert it into usable energy. While the concept has been a staple of science fiction and has drawn serious consideration from some scientists, I see no practical path to building such a structure.

Both Dyson spheres and Dyson swarms encounter overwhelming physical and engineering barriers. The material demands are staggering. Constructing even a thin shell just beyond

Earth's orbital distance would require dismantling entire planets to gather enough mass. This centuries-long endeavor would radically alter the mass distribution of the solar system, introducing long-term gravitational instabilities. Such perturbations could destabilize planetary orbits, including Earth's, leading to unpredictable and potentially catastrophic climatic consequences.

Structurally, the problems grow worse. The rigid shell of a Dyson sphere would require materials with tensile strengths far beyond the limits of atomic bonds. Even if we could manufacture such exotic matter, the gravitational dynamics are inherently unstable, demanding continuous energy expenditure for station keeping.

Dyson swarms are more physically plausible but come with their own headaches. The technical requirements for coordinating enormous numbers of independent units across astronomical distances may be a nonstarter. And let's not skip over the paradox of needing stellar-scale power to build a structure designed to collect stellar-scale power.

If a civilization were advanced enough to attempt a Dyson structure, it would likely already have access to stellar-scale energy through more efficient and elegant means. For example, controlled fusion or other exotic energy systems, such as matter-antimatter annihilation.

Tellingly, despite decades of sky surveys and the theoretical ease of detecting Dyson spheres via their waste heat signatures, we've found no compelling candidates across our cosmos. That absence suggests that Dyson megastructures are not a natural end point of technological evolution, but rather a conceptual dead end. Truly advanced civilizations would likely adopt energy strategies that work in harmony with the physical constraints of their environments, rather than against them, achieving similar power access with less environmental disruption.

Since 1952, humans have possessed the ability to create star-like energy through nuclear fusion—but only as brief, explosive events unsuitable for power generation. That's changing. In 2022, scientists achieved net-positive, controlled fusion energy output for the first time. The next step is sustaining it. Once we do, fusion will be refined, scaled, and adapted for a range of terrestrial and space-based uses.

Picture, for instance, a swarm of fusion-powered satellites orbiting Mars, Ganymede, or Titan and radiating light onto its surface. It's a vision that's both ironic and poetic. Ironic, because we humans once mistakenly believed the Sun orbited Earth. One day, we may live on worlds orbited by human-constructed artificial suns and peppered with fusion-powered solar radiator towers. Poetic, because the first major leap that set humans apart was the domestication of fire. Other animals use tools. Some communicate. But only we have mastered fire. By domesticating fusion energy, we will achieve the practical equivalent of a Type II civilization without dismantling our cosmic home.

Just for the sake of thoroughness, a Type III civilization harnesses the energy of an entire *galaxy*. I don't know what that Russian cat Kardashev was smoking, but that ain't happening either. Unless we figure out how to shrink whole galaxies down to size and transport them around, or figure out how to ascend the Marvel Universe and become Beyonders or Galactuses. Or is the plural Galacti? Only Stan Lee knows.

Back to our solar system. If (when) we ascend to a true Type II civilization that inhabits multiple bodies in our local area, then this brings us to another solution to Fermi's paradox, first encountered in the Realm of Life. Recall that Fermi's paradox asks, If there is other intelligent life out there, then why haven't we crossed paths with it? Maybe it's because the other supersmart life-forms out there realized they didn't need to go anywhere, even following the death of their precious star.

Until now, humans have imagined that the death of the Sun would mean that we would have to become an interstellar species and find a new Earth-like planet orbiting a new, suitable host star. However, the more practical solution is to stay within our solar system, powered by the stars we build ourselves.

Think of it this way: If we had the ability to build a warp drive or to sustain vast numbers of humans as we traverse empty, hostile interstellar space for millennia to reach Earth 2.0, wouldn't we also have the ability to build our own fusion-powered artificial stars? And if we could do that, then the question becomes, Why bother? We like it here, and we know from observations that our little corner of our cosmos is no different from any other suitable corner of our cosmos. This might be referred to as evolutionary isotropism: Across the universe, pretty much any relatively calm section of space will do. So let's stay put. Any self-respecting, technologically advanced, intelligent life-form out there will likely come to the same conclusion. And so, voilà! Science has answered Fermi's vexing question.

Ain't it fun to SWAG in the Realm of Imagination?

For humans, one billion years is a practical infinity (for our universe, however, it's a blip of a blip). During this time, Earth will undergo many calamities that humans must survive, including those of our own making, to sustain us for even one million years into the future. But it's gonna happen. Although it may be currently fashionable to take a pessimistic view of humanity's future, I remain hopeful.

Why? Because of our imaginations.

I stated early in this book that each realm has different rules, and that is no different for the Realm of Imagination. In fact, the only real rule here is that there are no rules. Whether it is possible to create a thing in the Middle Realm, in part by harnessing the powers inherent to the Quantum Realm, the Realm of Life, the Cosmological Realm, the Dark Realm, or any of the other realms, is beside the point. We must imagine it first, rules be damned. It could be a swarm of artificial stars or fusion reactors at sites all over planet Earth built by autonomous AI robots. Whatever it is, the first stop is the Realm of Imagination.

And I contend that our imaginations are here for the long haul. At this stage, it's an evolutionary imperative. That may offer another answer to our big question: Why do we exist? To stick around for an effective eternity, that's why. Heck, if sharks and scorpions and gators and dinosaur-birds can hang around for hundreds of millions of years, why can't we? Our longest-lasting homo ancestors, *Homo erectus*, lasted for two million years, domesticated fire, doubled their brain volumes, and likely invented language. Certainly we sapiens can outdo those hippies!

My realists out there, or perhaps my "cynics," might now be screaming, "But what about the end of the world, Dr. O.?!" Well, I'd like to disabuse you of the commonly held fallacies of impending human extinction. It's time for a little more deprogramming. Our normal deception will not do.

Nuclear war, global climate change, pandemics, and biodiversity loss represent the typical list of self-inflicted existential threats. In five years or so (or five weeks for my super techno-pessimists), we may add AI takeover to the list, but we're not there quite yet. The point is, we constantly devise new ways to end the world. While the occurrence of any of these would be devastating, none are likely to completely wipe out the entire human species and all our accumulated knowledge. Let's address them in turn.

The existential threat from nuclear war is the so-called nuclear winter. Yes, millions would die in explosions, but the aftermath is even more troubling. (Most sensible people would pick instant annihilation over postapocalyptic survival.) Nuclear winter assumes that many nuclear explosions above Earth's surface would ignite wide-ranging firestorms that would send large amounts of ash and soot into the upper atmosphere. This soot would absorb sunlight, preventing it from reaching Earth's surface. If this condition persists for years, there will be two primary effects: Photosynthesis will cease, and the food web will collapse. The other problem is related to the fact that Earth's atmosphere is heated by infrared light emitted from below, not the visible light emitted from above. The Sun's visible light heats the ground and oceans, which reradiate some of that energy as infrared light, which heats the atmosphere. If the Sun's visible light is unable to make it to Earth's surface, then there may be considerable cooling of Earth's atmosphere. Thus the phrase "nuclear winter."

There are two primary ways to evaluate the threat of extinction from nuclear winter. The first is to perform the experiment—loft soot into the upper atmosphere and see what happens. The second is to computationally simulate the same scenario. Both have been done extensively.

The meteoritic impact that wiped out the dinosaurs was the explosive equivalent of one hundred billion atomic bombs. Following the impact, firestorms raged across the planet. The atmosphere was filled with ejecta, gases, and dust. Gases can remain airborne for millennia, but ejecta and most soot and ash, which are composed of solids, eventually fall back to Earth. Earth's skies darkened for somewhere between ten and twenty years sixty-six million years ago. The major culprit was fine dust in the ejecta, not the ash and soot, because it remained aloft, suspended in the upper atmosphere, and absorbed sunlight for around fifteen years.

Many sources have recently injected Earth's atmosphere with gases and aerosols. The closest analogue to hundreds or thousands of nuclear explosions occurring quickly and triggering firestorms may be the massive wildfires of 2023 and 2024. In the Southern Hemisphere, Australian bushfires burned around 371,000 square miles, which is slightly larger than the combined areas of Montana, Nevada, and Oregon. In the Northern Hemisphere, Canadian wildfires in 2023 burned more than 71,000 square miles, an area equivalent to that of the state of Washington. Yet 2023 and 2024 were the hottest years on record.

By comparison, the Hiroshima nuclear bomb detonation burned an area of only around six square miles. It would take sixty-two thousand of those to equal the area burned by the Australian fires. It's currently estimated that the world possesses fewer than fifteen thousand nuclear bombs in total. And we're certainly nowhere near having one hundred billion to reproduce the deadly meteor that struck Earth sixty-six million years ago. Life survived that, and life back then was dumb as hell!

So yes, nuclear war would be terrifying and paradigm shifting, but from an evolutionary standpoint nuclear winter ain't nothing to fret about.

And believe it or not, the same can probably be said of climate change. Consider this quotation from MIT's Adam Schlosser: "If I had to rate odds, I would say the chances of climate change driving us to the point of human extinction are very low, if not zero." This is from the Deputy Director at the MIT Center for Sustainability Science and Strategy. Just as with nuclear war, most significant impacts will be regional.

Will surviving and adapting to either of these realities be fun or easy? Nope. Tens of millions or more will perish, and surviving will be extremely difficult and unpleasant. But either one causing human extinction is unlikely.

Pandemics and biodiversity loss may also wreak havoc on civilization and human lives. But history demonstrates that future challenges will be met with the same creativity, cooperation, and determination that have defined our species. The rapid development of mRNA vaccines during the COVID-19 pandemic exemplifies how science can respond to an urgent biological threat. And the tools we have at our disposal here will only improve.

Regarding pandemics, future humans will continue to refine our knowledge of pathogens and their transmission modes. We'll develop sophisticated surveillance systems that detect outbreaks before they spiral out of control, using artificial intelligence to perform real-time analyses of global health data. We will perfect the technologies of universal vaccines, develop antivirals that can be taken pre-exposure, and maybe even genetically engineer the immune system to resist infections. If reality truly follows art, we may eventually all have a parallel nanobot immune system that works in concert with our natural immune systems.

Human resilience in the face of biodiversity loss will also rely on a mix of technological innovation and systemic change. Biotechnology will play a pivotal role, utilizing techniques such as gene editing to restore species and ecosystems. "De-extinction" and rewilding efforts may bring back species crucial to stabilizing ecosystems, while genetic modifications could enhance organisms' ability to rapidly adapt to changing environmental conditions. Simultaneously, humanity could integrate regenerative practices into agriculture, urban design, and energy production to minimize the ecological footprint of modern society (looking at you, fusion!).

As wacky as it may sound, Jeff Bezos's company Blue Origin envisions a future where heavy industry, including manufacturing and energy-intensive processes, as well as large-scale agriculture, is moved off Earth and into space. In this scenario, Earth can be treated like a kind of wild garden and allowed to return to the Eden it once was.

Okay, cue my realist-cynics again. "But look at the world right now! How can you be such a Pollyanna, Dr. O.?" If you flip back to the Cosmological Realm chapter, you'll see that I wrote that size matters. There I was talking about the vast scales of our cosmos and how this vastness relates to things like gravity wells and galaxy formation. A corollary here is that time matters. As in passage of time. If you are despairing in this moment—or any moment—just remember that nothing lasts. Things change. The Temporal Realm revealed that our universe has hundreds of trillions of years to go. The Quantum Realm showed us that possibility and outright weirdness are a kind of observable magic that lies at the root of all creation. The Dark Realm showed us that invisible forces govern our very existence. The Realm of

Life showed that our cells and biological mechanisms are feisty as hell. The Multiverse Realm revealed that all realities are not just possible but guaranteed, that there is no such thing as fiction. And so too for us: In the Realm of Imagination, there is no such thing as a limitation. As long as we persist, we will overcome whatever barriers come our way. And I am certain we will persist for a long, long time.

How long, you wonder? We can examine humanity's deep future at three milestones: one million, ten million, and one hundred million years into the future.

The first waypoint, at one million years, is comparable to the reign of *Homo erectus*. 'Nuff said. We got that. *Homo sapiens* have been around for three hundred thousand years or so already. We'll outlast Stone Age humans. They knapped rocks into cutting tools. We melt sand into thinking tools.

The next benchmark is more of a natural species lifespan horizon because most species do not survive beyond ten million years. This is known as the background extinction rate. It reflects the natural cycle of evolution and species turnover. Over time, environmental changes such as shifts in climate and habitats compel species to adapt or face extinction. In some cases, species evolve into new forms, effectively marking the end of the original species (see: dinosaurs/birds).

Our genus, *Homo*, has undergone rapid evolutionary changes throughout its history. In relative terms, significant developments have occurred over a relatively short geological period. This evolution is characterized by rapid increases in brain size, changes in our physical bodies, and notable behavioral adaptations.

The earliest known representative of the genus, *Homo habilis*, appeared around 2.8 million years ago and was the first to exhibit evidence of stone tool use. This marked a crucial behavioral shift. Around 1.8 million years ago, *Homo erectus* emerged, marking further advancements, including the controlled use of fire and the development of more complex tools. This species was also the first to expand beyond Africa, dispersing across Asia and Europe. Modern *Homo sapiens* originated within the past three hundred thousand years in Africa and have spread to every corner of the planet.

Humanity's future evolution, which will integrate technology into our bodies and brains, promises to blur the boundaries between the natural and the artificial. It will be a profound transformation, driven not by the slow, imperceptible march of natural selection but by deliberate design and innovation. With the convergence of biotechnology, artificial intelligence, and nanotechnology, humans may gradually become cyborg-like entities, enhancing our physical, cognitive, and emotional capacities in ways that were once the domain of science fiction.

Initially, this evolution may concentrate on overcoming our biological limitations. Neural interfaces could enhance memory, intelligence, and creativity, allowing us to process and access information at currently unimaginable speeds. These technologies might foster new forms of communication, enabling the direct sharing of thoughts and emotions and bypassing the constraints of language altogether. This more profound interconnectedness could redefine human relationships and society, creating networks of minds linked through a shared digital consciousness.

Integrating more technology into our bodies could also change our physical forms. Prosthetics and exoskeletons may replace or enhance limbs to restore lost function and exceed it. We will possess strength and endurance beyond our natural capabilities. Internal devices could regulate our health in real time, monitoring and repairing cells, eliminating disease, and potentially extending our lifespans indefinitely. These changes may usher in a new era where aging is no longer a fixed trajectory but a modifiable aspect of existence.

Yet, as we embark on this path, we will likely find ourselves revisiting the most profound questions of our existence. What will remain of the essence of being human as we evolve into something fundamentally different? The answer is obvious. Even without technology, humans one million years from now will be mentally and physically different from today. Humanity's future evolution, therefore, is not just a matter of technological advancement but a journey into the heart of what it means to exist, adapt, and thrive in a universe of our own making—a universe, as it were, of our own imagining.

One of the most iconic and influential images of human evolution is the unfortunate "March of Progress" visual, depicting humans evolving from an apelike ancestor over twenty-five million years. It started innocently enough in 1865 when an author placed a human skeleton last in a left-to-right sequence of other great ape skeletons. The order was gibbon, orangutan, chimpanzee, gorilla, and human. It was designed to illustrate the similarities between humans and other great apes, uniting us with the animal kingdom; it was not meant to be a depiction of our evolution. However, readers interpreted it as illustrating monkey-to-man linear progress evolution. The meme stuck, and one hundred years later it showed up in its two current forms, originally titled "The Road to Homo Sapiens," in Time-Life's twenty-five-volume Life Nature Library.

There were two illustrations in this Time-Life series: One contained fifteen primates, and the other contained a subsample of only six. In the first illustration, a third of the primates were not human ancestors. In the latter and more popular version, the last two are essentially the same human species, and the four that precede them are extinct non-human-ancestor apes. Unfortunate indeed!

In 2023, a book titled *A Brief History of Intelligence: Evolution, AI, and the Five Breakthroughs That Made Our Brains*, by Max Bennett, presented what amounts to a road to *Homo sapiens'* intelligence and imagination. This is our actual evolutionary trajectory as primates. It is grounded in the idea that as life evolved, so did the complexity of the brain and its capacity for increasingly sophisticated cognitive abilities.

The journey begins with simple life-forms that possess only basic neurological structures, allowing them to respond to their environment. Picture the first multicelled organisms drifting around in primordial seas, like those we met in the Realm of Life. Then came a revolutionary change: the emergence of a rudimentary nervous system. This wasn't yet a brain but a basic network of neurons that enabled creatures to sense and react. Imagine a jellyfish pulsing through the water, guided by its nerve net toward food and away from danger. Over time, evolutionary pressures led to the development of more advanced neural mechanisms, allowing organisms to interact with their surroundings in increasingly complex ways.

Millions of years later, long after nervous systems had evolved into brains, a game-changing breakthrough occurred—the ability to learn from experience through reinforcement. This leap was driven mainly by two critical components of the brain: the midbrain dopamine system and the basal ganglia. The midbrain dopamine system evolved into a sophisticated reward predictor, functioning like a tiny fortune teller constantly updating its expectations. When something unexpectedly good happened, it released a burst of dopamine, signaling, "That was dope!" (Pun intended.)

Meanwhile, the basal ganglia, particularly a region called the striatum, became the brain's action selector. Think of it as a control room operator, choosing behaviors based on past rewards. Together, these systems enabled our ancestors to identify which berries were safe to eat and which predators to avoid, thereby vastly improving their chances of survival.

The next major leap was the ability to predict outcomes without direct action. Instead of relying solely on trial and error, organisms could envision different possibilities and choose the most advantageous path. This is where the neocortex—the wrinkled outer layer of the brain—became crucial. Early mammals had a relatively simple three-layered cortex, but over time it evolved into a complex six-layered structure, capable of processing information in incredibly sophisticated ways.

With this growing cognitive complexity came the ability to infer the thoughts, emotions, and intentions of others. A skill known as theory of mind. This metacognitive leap likely involved further refinement of the prefrontal cortex, particularly to regions associated with self-awareness. Another key player in this development is thought to be the angular gyrus, a hub that integrates different types of information. It plays a crucial role in constructing a coherent sense of self.

Understanding others' intentions and predicting their behaviors became essential for cooperation, competition, and survival within groups. This laid the foundation for rich social relationships and, ultimately, one of the most significant cognitive leaps: the development of language and the ability to record thoughts in writing.

Human language has profoundly transformed the brain. It led to hemispheric specialization, with language functions primarily concentrated in the left hemisphere. Two regions became particularly important: Broca's area, in the frontal lobe, which governs speech production, and Wernicke's area, in the temporal lobe, which enables language comprehension. The temporo-parietal junction, which includes Wernicke's area, also expanded significantly. This region acts like a linguistic Swiss Army knife, supporting various aspects of language processing and social cognition.

Language didn't just allow humans to communicate; it enabled the creation of shared realities and cultural frameworks that extend beyond individual experience. The ability to exchange ideas, preserve knowledge, and build complex societies is a hallmark of biological intelligence.

Throughout this evolutionary journey, our brains have grown larger and become more densely packed with neurons. The cortex became increasingly folded—like a crumpled piece of paper—maximizing surface area within the confines of the skull. This expansion provided the neural capacity for even more complex thought processes.

In humans, the prefrontal cortex, located just behind the forehead, expanded dramatically. One of the most uniquely human cognitive abilities to emerge is prefrontal synthesis (PFS)—the ability to consciously combine objects, ideas, or events into novel mental scenarios. Closely linked to the advanced functions of the prefrontal cortex, PFS enables us to imagine, plan, and solve problems in ways that distinguish us from other animals. It enables us to mentally "play" with ideas, creating new combinations that don't exist in reality.

While many animals exhibit impressive cognitive and language skills, their thinking is often tied to immediate experiences or fixed patterns of learned behavior. A chimpanzee, for example, may use a stick to extract termites from a mound, but it doesn't spontaneously invent a completely new tool to get more termites faster. Humans, thanks to PFS, can envision solutions far beyond their direct experience—an ability that fuels invention and innovation.

It is the Realm of Imagination, and it is ours alone.

(At least in this corner of our universe.)

▪ ▪ ▪

On our journey through the Nine Realms, we have traveled from the familiar to the weird, through the mysterious, and into the speculative. Each realm reveals something essential about our universe—but none more so than this one: the Realm of Imagination. The jewel of human intelligence. A product of our cosmos. And the only lens through which our cosmos can understand itself.

Every life-altering scenario described previously—nuclear winter, climate change, biodiversity loss, the death of the Sun and Earth—will present us with conditions that appear hopeless. But our imaginations, which have evolved over millions of years and will continue to evolve over millions more, will guide us. Our imaginations give us hope. They also allow us to take action.

No outcome occurs without first imagining it. No theory of the universe is proven without first observing the world and then using our imaginations to devise that theory. Stars will come and go, but we have the capacity to make our imaginations live essentially forever.

On this barely-there speck of dust in the vastness of the Nine Realms, consciousness arose. Our universe—its stars, planets, galaxies, plasmas, nebulae, black holes, fields, quanta, and energy—works mindlessly to create minds. *Our* minds. How cool is that? We emerged from fire and chaos—literally—and we will figure out ways to meet what comes next. If our universe, so brutally indifferent to life, still gave birth to thought, then that alone is cause for hope.

We are the only creatures on Earth that understand the reality of stars. Life knows the Sun's warmth, but only we know its fusion, its fields, or its fate. Only we know it has trillions of cousins, scattered across time and everywhere at once. The stars are the little engines of our universe. Our imaginations—nebulous, barely understandable expressions of energy transfer—are, perhaps, the stars' greatest creation. The evidence suggests this: The universe and imagination evolved together.

Some will call this view techno-humanist and criticize it for ignoring constraints, power structures, or the damage we've already done. Fair. Imagination and reason gave us nuclear weapons and climate collapse, too. Why trust them now?

To some, techno-humanism may sound like arrogance masquerading as optimism. To others, it displaces God with gadgetry. It's seen as a hollow gospel of neural implants and fusion reactors, promising salvation through science.

I don't claim this view is the only one. Many find meaning in spiritual connection, religious faith, or philosophical inquiry. I respect those paths. But I also find that faith in the human mind, our capacity to wonder, to build, to change, is not incompatible with reverence. It is a form of reverence.

Is it all a dream, as D'Angelo (or Biggie) might ask? No. Would our universe exist if we or some other alien consciousness didn't perceive it? Probably yes. But without us—or something like us—it would not be *known*. It would not be *imagined*. And so, it would not be this universe. Not the one we live in. Not the one we shape.

We cannot have a better future to work toward if we lack the ability to first think it into being. And we astro-folks need your help, even if you've never opened a physics textbook or taken calculus or scanned radio telescope data. We need you to guide us, push us, fund us, help us, encourage us.

You are a being of infinite complexity, defined by relationships—relationships between atoms, molecules, and fields, as well as those relationships between your experiences and the world around you. You are an emergent phenomenon, shaped by the interaction of countless systems: physical, biological, and cultural.

At your core, you are a question in motion—a dynamic process rather than a static object. You are possibility, curiosity, and change embodied, continually constructing and reconstructing yourself as you engage with the mysteries of existence. What you are defies simple categorization because you are always in a state of becoming. Sitting at the heart of all that is your imagination.

This book may be naive. Or arrogant. Maybe both. But history shows that where collapse seemed inevitable, we've often found a way forward. Not always wisely. Not always justly. But often enough to survive. If we are to do more than survive, if we are to flourish into deep time, we must do what no other species has done: dream at scale and build with care.

Imagination, more than any other human trait, is the fundamental precondition for progress.

It is the excitation in the field of possibility.

ACKNOWLEDGMENTS

This book is the product of far more than my own labor. Writing about the universe is not a solitary act—it's a conversation across time, space, and community. I owe a debt to those who challenged me, supported me, and reminded me that knowledge only matters when it's shared.

First, to my family: Your patience and love kept me steady when my mind was elsewhere—lost in equations, chasing metaphors, or orbiting too far from Earth. You gave me the grounding I needed to complete this work.

To the many thought partners and early readers who shaped my growth as a scientist and sharpened my ideas: Richard McGinnis, Dave Teal, Gerald Bruno, David Santiago, Alex Kim, Eric Linder, Katherine Scott, Darin Brown, Art Walker, Holly Shah, Kervin Evans, Michael Levi, Kayla Violet, Ed Ratner, my old MSSTA crew, my NASA colleagues, and the countless students who forced me to explain things more clearly than I thought possible. You pushed me to think harder, test assumptions, and pursue rigor with respect.

To my editor, agent, and publishing team: Thank you for helping me shape raw thoughts into a narrative form. Every cut, comment, and question made this book sharper and more alive. A special thanks to those who endured my late-night bursts of revision and my stubborn attachment to pet phrases that didn't deserve saving.

Finally, to the readers: You are the true co-authors. A book like this comes alive only in the act of reading, in the spark of recognition or wonder it stirs in you. If, for even a moment, these pages make you pause, look up at the night sky, and feel the vastness that makes us possible, then this work has done its job.

IMAGE CREDITS

INDEX

Note: Page numbers followed by *f* indicate figures.

ABOUT THE AUTHOR

DR. HAKEEM OLUSEYI, author of the critically acclaimed memoir *A Quantum Life: My Unlikely Journey from the Street to the Stars*, is a multidisciplinary astrophysicist, multi-patented inventor, award-winning author and journalist, internationally recognized educator, and highly sought-after TV presenter, podcast host, voice actor, and keynote speaker. He currently serves as CEO of America's first and longest-standing national astronomy organization, the ASP (astrosociety.org).

X: @HakeemOluseyi